What a Wonderful World

Presented By
the jazz club
of Sarasota
366-1552

WHAT A WONDERFUL WORLD

A Lifetime of Recordings

BOB THIELE
as told to
BOB GOLDEN

Foreword by Steve Allen

New York   Oxford
OXFORD UNIVERSITY PRESS
1995

Oxford University Press

Oxford   New York   Toronto
Delhi   Bombay   Calcutta   Madras   Karachi
Kuala Lumpur   Singapore   Hong Kong   Tokyo
Nairobi   Dar es Salaam   Cape Town
Melbourne   Auckland   Madrid

and associated companies in
Berlin   Ibadan

Published by Oxford University Press, Inc.,
200 Madison Avenue, New York, New York 10016

Oxford is a registered trademark of Oxford University Press

Library of Congress Cataloging-in-Publication Data
Thiele, Bob.
What a wonderful world : a lifetime of recordings /
Bob Thiele ; as told to Bob Golden ;
foreword by Steve Allen.
p.   cm.   Includes index.
ISBN 0-19-508629-5
1. Thiele, Bob.
2. Sound recording executives and producers-United States—Biography.
I. Golden, Bob.
II. Title.
ML429.T44A3   1995   780'.92—dc20   [B]   94-23125

9 8 7 6 5 4 3 2 1

Printed in the United States of America
on acid-free paper

To my wife Teresa,
my son Robert Jr.,
and Duke Ellington

Foreword

By Steve Allen

Jazz has at least certain factors in common with religious faith. Its followers may be fanatically devoted, they adhere to their views even if they are not widely shared, and they will go to considerable trouble to attend ritualistic ceremonies. There is one particular, however, that jazz does not share with the thousands of religious denominations in history, and that is that no lover of jazz has ever decided, after thinking things over, that his faith was misplaced. Once you develop a true understanding of jazz you may be assured that the appetite for it will last as long as you do.

Jazz also shares with religions the tendency to endless heretical deviations from the status quo. It fortunately does not have in common with religion the tendency to subject heretics to aggressive attack, including the physical and the military.

It is, in any event, the only art form America ever created and as such has generally been accorded appropriate cultural respect. Unfortunately it has, especially in recent decades, not always been favored with the proper degree of popularity. Things were better in this regard during the Golden Age of American popular songwriting, that glorious 30-year period—the 1920s, '30s and '40s—in which not only America's best composers gave us the fruits of their astonishing creativity, but players who are still recognized as giants provided a large volume of music in the jazz idiom. The great bulk of this, of course, was as evanescent as conversation itself, but, fortunately, the introduction of the technology of the recording industry made it possible to capture, for all time, moments of blazing achievement.

Of course none of that happened simply because separate groups of musicians accidentally happened to physically encounter open microphones that led to recording tables. The records happened because certain executives planned it so, either by bringing their equipment into jazz clubs and theaters, or, as in most instances, inviting the musicians into recording studios. The performers had two motives for taking part in the largely happy process: a) they were paid for their services; and b) their egos perceived the general rightness of having their work preserved for the ages. The record industry people also were motivated by two considerations: first, a perception of the wisdom of preserving the performances of gifted artists, and, second, the awareness that one could make a buck by doing such a nice and sensible thing.

At the larger recording companies, only some of the product introduced into the marketplace consisted of jazz. A problem arose when it became apparent that many of the nonjazz forms were far more profitable. We are fortunate indeed, as a people, that the corporate thirst for profit, which is certainly not intrinsically evil, did not dictate all decisions. There were certain men behind executive desks who were motivated more by their simple love of jazz and its performing representatives than by purely economic considerations. Such a one—and may their tribe increase—is Bob Thiele.

Given that the great majority of jazz recordings, even by major artists, make very little money, it was, of course, necessary for Thiele's various companies to appeal to broader segments of musical taste, with the result that he has been importantly instrumental in furthering the careers of the McGuire Sisters, Henry Mancini, Eydie Gormé, Steve Lawrence, Teresa Brewer, and your obedient servant, to name only a few. Almost all my important early single albums were done under his auspices, and I can not recall a single instance in which there was the slightest difference of opinion between us. If he had an idea for an album I would simply accept the assignment. If I had a suggestion he seemed invariably to find it reasonable.

In the field of rock he discovered Buddy Holly and Jackie Wilson and produced their first hit records. He was also wise enough to bring such blues specialists as B.B. King, John Lee Hooker, Otis Spann, and Joe Turner to new levels of popularity.

The thought occurred to me some years ago that no one can as fully appreciate a given art form as those who are practitioners of it. The lay public, for example, obviously grasps the greatness of, say, Oscar Peterson, Phineas Newborn or Erroll Garner. But I believe it is only their fellow pianists who fully apprehend the beauty and wonder that such artists are capable of creating. As for Mr. Thiele, it is probably the fact that he started out as a musician himself that explains his unusually sensitive understanding of what jazz musicians do. Just so I believe that it is the fact that Bob has tried his hand at songwriting that has always made him particularly sensitive to the genius of Jerome Kern, George Gershwin, Duke Ellington, and other Golden Age masters of the art of melody and lyric writing.

Well, enough of this; you can read what I have to say anytime. But now you should read this fascinating recollection of a productive life spent in the music business. Its individual anecdotes and reminiscences are priceless, and one of them, an account of a bizarre party involving a number of American recording executives and a group of their German counterparts, representing Deutsche Grammophon, is by itself worth the price of the book.

Contents

Introduction

Every Mardi Gras season, the city of New Orleans is spectacularly transformed from a metropolis of indifferent strangers into a friendly village of neighbors. Since my wife Teresa Brewer and I needed a holiday after we completed several weeks of intensive recording for my RCA-distributed Flying Dutchman label, a vacation in New Orleans seemed most timely.

During lunch in the hotel dining room the day after we arrived, we heard through the windows a band out in the street playing Dixieland that was even more delicious than our meal, so we quickly joined the crowd that had gathered around these enthusiastically masterful musicians. (We later found out that the drummer was the grandson of jazz legend Paul Barbarin, who played drums with Louis Armstrong before Satchmo and jazz went up the Mississippi.)

The band leader was Scott Hill, a veteran Crescent City trombonist who immediately recognized Teresa among the spectators and said, "Come on, sing with us," an invitation she, thoroughly in tune with the festive occasion, happily accepted. At the same time, a photographer who was a friend of ours, with an apartment balcony that overlooked the street, offered me his hospitality and the vantage to observe the merriment below.

With drink in hand and contentedly perched above several hundred ecstatic celebrants, I continued to watch the revelry and hear my wife, without a microphone, sing one Dixieland classic after another. On her third chorus of "When the Saints Go Marching In," another dear friend emerged from the throng into my view. It was Toby Pieniek, an attorney with RCA Records in New York, who was in town for a corporate convention.

Seeing me on the balcony, he shouted up, "Bob, you're the greatest promotion man in the record business. What a great idea to have Teresa Brewer on Bourbon Street with this crowd. You masterminded this event perfectly!"

At that exquisitely ambrosial moment, my first thought was, "This crazy business I'm in has even followed me down to the Gulf of Mexico." And then, "How lucky I am to be part of the record industry, with good friends in this fantastic place listening to Teresa, my favorite singer and the best friend I'll ever know."

There we were, sharing the music I loved with all those joyous people. It couldn't get any better than this. What a wonderful world to live in!

What a Wonderful World

1 Newton's Law

THE song that shares its title with this book had an unnaturally troubled genesis.

My most revered idols as a kid growing up were Louis Armstrong and, of course, Duke Ellington. When I got into the record business, I dreamed of the day I could record Louis Armstrong, and the first recordings I made with him were for Morris Levy's Roulette label. Louis even improved on the production idea and made my dream come truer: he brought in his little band—five pieces—and agreed to use a friend of mine on piano, Duke Ellington!

Some years later, in the mid-1960s during the deepening national traumas of the Kennedy assassination, Vietnam, racial strife, and turmoil everywhere, my co-writer George David Weiss and I had an idea to write a "different" song specifically for Louis Armstrong that would be called "What a Wonderful World." We wanted this immortal musician and performer to say, as only he could, the world really *is* great: full of the love and sharing people make possible for themselves and each other every day.

At the time Louis's "Hello Dolly" was the biggest hit record in the country and, as a result, Armstrong was a bigger star than at any previous moment in his career. As he was constantly on tour, and we were anxious for our new song to be approved in time to follow his current hit, I went to Joe Glazer, who was Louis's manager and, as was his client, a dear friend of mine, to ask if I could visit Louis on the road and make a presentation to him.

With Glazer's permission, and a small children's portable phonograph to play the acetate disc recording George and I had cut of the song, I went down to Washington, D.C., where Louis was appearing. Between shows I auditioned our number for him on my toy record player, and, before he had listened to it the first time through, Armstrong said, "Pops, I dig it. Let's do it!" (Of course, we called him "Pops," and he called *everybody* "Pops.")

We scheduled the recording session in New York City after Armstrong's Washington engagement. This was to be Louis's next record after "Hello Dolly," which was a typically up-tempo Dixieland-style song Armstrong was popularly famous for. As a ballad with violins, cellos, and a rhythm section, "What a Wonderful World" would be (we hoped) an attractive departure for Louis and his fans. In fact, Louis agreed to record it for minimum union scale (approximately $250 at the time) because he liked both the song and this new concept for him, and was mindful of the expense required for the extra string musicians to achieve the desired effect we envisioned.

The then president of the record company, ABC Records (where I was also employed as an executive), was a crudely arrogant and volatile corporate climber named Larry Newton. When he showed up at the recording session, his first concern was to find out if a photographer was present, because he wanted a picture taken with Armstrong. After the photo was taken, and Newton heard the music as the session started, he became visibly upset and screamed at me that I had to be crazy to record a ballad with strings as the follow-up to "Hello Dolly" rather than another fast Dixieland tune.

Although I attempted to explain that that was the idea (and why it had appealed to Louis and Joe Glazer)—we would reverse the prevailing industry belief (and the inevitably diminishing sales) that Armstrong could be successful only with the same thing again and again—Newton was only temporarily pacified. As the recording progressed, he became increasingly incensed and disruptive as his agitation about this "radical" concept intensified. Finally he declared he wanted to cancel the date and fire the musicians and me as well.

I was already a gelatinoid wreck from my efforts to keep my boss quiet and not have my star and friend become as nervous as I was. Now, all I could do to convince Newton to momentarily leave the control room and reconsider was to tell him he would go down in history as the only man who ever threw Louis Armstrong out of a recording studio.

This tenuous calm ended almost immediately when a freshly wrathful Newton tried to storm back into the studio. Frank Military, one of our friends who was also the music publisher of our beleaguered ballad, literally became a human barricade at the entrance so Newton could not burst in. The ensuing door-pounding by my now ballistic boss caused Frank, an unflinchingly hard-nosed veteran of numerous music business wars, to actually begin crying and plead, "You can't do this to Louis and Bob." Miraculously, with all the sinister drama and ominous distractions, the recording of one of the most optimistic songs ever written was completed.

When these previous events were recounted to Joe Glazer the next day, he called Newton and offered to personally buy back the recording regardless of its cost. A petulant Newton refused, and a rapidly escalating feud developed between them.

At a sales meeting when "What a Wonderful World" was to be released as a single, Newton couldn't wait to ridicule the record when it was played for the assembled executives. "Wait till you hear this stupid record," he said and added, glaring directly at me, "made by my *Nazi* friend Bob Thiele!" Predictably, the meeting erupted into a riot as I exploded out of the conference room with declarations of where they could shove the record company and, while they were at it, my contract, swiftly exiting the building and loudly quitting my job to anyone within earshot. (I was followed down Sixth Avenue by a vice president, David Berger, who implored me to return—I was making a big mistake, etc., etc.—and, a few days later and after more strident dramatics, I was convinced to be back at work.)

In a few weeks, the record did come out in the United States and, personally sabotaged by a bitterly vengeful Newton, was a disaster. If a thousand singles had been bought, it would have been a lot—and this right after the million-plus units "Dolly" had sold. In

England however, it became #1 with sales of over 600,000 copies and remained #1 for the thirteen weeks it outsold both the Rolling Stones and the Beatles. It also started to become a hit in many of the other European countries and, to everyone's amazement, South Africa. The EMI Corporation, which licensed ABC Records for foreign distribution, sent a telegram to Larry Newton saying MUST HAVE WHAT A WONDERFUL WORLD ALBUM, which meant we needed to record eight more songs in the same style to have a complete album.

Newton couldn't stand it. He told me the hell with it, he wouldn't ever again deal with that ingrate Joe Glazer and didn't care about the European companies. And if this was so important, it was now *my* responsibility to speak with Joe and get Armstrong to record the additional songs at the original rate for a total of $500. An equally enraged Joe Glazer advised me to forget it unless Newton paid Louis Armstrong a then undreamed-of $25,000 advance for the eight more sides.

I was the besieged broker between two rabidly adamant antagonists. (Glazer: "I heard what went on. You tell that fat bastard to go fuck himself and give us twenty five thousand dollars for eight more sides." Newton: "Tell him to go fuck himself and why do we give a shit about those European companies? Screw 'em all.") This went on and on as urgent telegrams, cables, and telexes persistently poured in from all over the world begging for a full "What a Wonderful World" Armstrong album.

When the pressure from overseas finally caused Newton to relent and approve the $25,000 payment (he would always call it a ransom), Louis Armstrong completed one of the best-selling albums of his lifetime and again, posthumously, when the original single became the centerpiece of the soundtrack for the mega-hit movie *Good Morning Vietnam* a quarter-century afterwards.

As for Larry Newton, his executives at subsidiaries around the globe and his corporate superiors at home applauded him for an innovative and astoundingly successful use of Armstrong's talents. But he never said a word of appreciation to any of us who created the song and the record. I believe the episode represented a deeply personal defeat which privately continued to gnaw at Newton's injured ego as both the single and album became one of the most enormous triumphs he and his record company would ever achieve.

Youth and
the Night
and the Music

2

I

AS Brooklyn already had the Dodgers, Coney Island, and a few blocks from our home in Sheepshead Bay, our family-owned (and famous) Lundy's Restaurant, my arrival on July 27, 1922, was as uneventful as the childhood that ensued. I went—actually could walk from my house on East 21st Street and Vorhees Avenue—to school at P.S. 98. Lundy's Restaurant was operated by Uncle Irving, my mother's brother, Frederick William Irving Lundy. Mom's maiden name was Athenaise, and Herbert Thiele, extolled by my relatives far and (increasingly) wide as "one of the great bulk chocolate salesmen," was my father, whose accounts included the Fanny Farmer and Barraicini companies in the East, Mary Oliver Candies in Chicago, and the Sees Co. in Los Angeles.

Through several generations, my family became established as Sheepshead Bay aristocracy. Frederick Lundy, my brawling, boisterous grandfather, even ran for a seat on the Brooklyn City Council. (Without success, as the principal New York City newspaper at the time, *The Journal-American,* delighted in their reportage of Grandfather's frequent nocturnal antics. It was not unusual to read about his visits to bars throughout the borough where he would buy drinks for everyone present, or about the inevitable fistfight that he provoked when his "guests" did not entirely agree with what he was saying and didn't indicate enough appreciation for his generosity. Once, the *Journal* gleefully called attention to a costly lawsuit he lost as a result of his drunken brawls. The amount involved was $5000—an unimaginably monumental fortune in those days.)

Around 1930, when I was eight, my parents decided to move to the "better environment" Forest Hills (in the borough of Queens) offered. Although no interest in music was evident during these years, I recall my father seemed to enjoy the 78 rpm records which he would bring home for the family to hear with him.

I attended P.S. 101 in Forest Hills Gardens and led—at the time—a blandly typical pre-teenage life of the son from a well-to-do family who could come home for lunch (usually a peanut butter sandwich and milk), and where the front door was always open—a style of living nonexistent today. Since we lived a few blocks from the old West Side Tennis Club, I somehow was drawn to the tennis stars who played there. I was never interested in the actual sport, but, so enthralled with the glamour and excitement of the games, I wound up parking cars and selling seat cushions for spectators at the stadium. In fact, whenever a tournament was played and my father had driven home from his office, he inevitably would have to find a parking space—usually at a great and time-consuming distance from our house—because the garage and driveway of the Thiele residence was fender-to-fender with the automobiles of my valued customers.

My attraction to a life of fabulous glamour was further ignited a few years later when I was thirteen, and my father took me with him to California when he made his annual visit to his big candy manufacturing accounts on the West Coast. By prearrangement, we stayed at the mansion on Sunset Boulevard in Bel Air of my two cousins Douglas and George "Buddy" Ornstein, who had inherited (and, with blinding rapidity, subsequently squandered) vast wealth from their father, who had married my dad's sister, Elsie.

I was gleefully introduced and induced into the fabled 1940s royal and rowdy Hollywood Life Style of the impossibly rich and frivolous with a party every night at this movieland manor: poolside bacchanals until dawn, nude beautiful people everywhere, everyone bombed. (My bedroom looked out over the swimming pool, and I could literally jump into the water from my window—which was how I woke up every morning just before noon. Talk about falling out of bed!)

I absolutely worshiped my cousin Buddy and was his gratefully constant companion. His girlfriend was Gwen Pickford, whose late mother was the sister of Mary Pickford, then married to Douglas Fairbanks and, with Charlie Chaplin, the movie superstar owners of the original United Artists film studio. As Gwen was now the adopted daughter of Mary and Douglas, we would all spend many a halcyon afternoon at the storied Pickfair estate, swimming and drinking. I especially enjoyed pressing the button in the spacious main living room at Pickfair and watching a wall disappear to reveal the entire (and fully stocked and furnished) western saloon that Mary had had built for her husband as a surprise birthday gift.

Buddy was also among the naturally funniest people I had ever known. He had already achieved some legend as a "star" test pilot for Hughes Aircraft who once landed a prized experimental plane upside down and exited the apparatus by falling out on his head. He never flew again.

Years later, he and Gwen got married, and he was appointed the chief of operations for United Artists in Spain. From everything I heard, no ambassador anywhere mixed or fixed more lavishly than my old buddy Buddy. He was a frequent, honored guest at the palace of Generalissimo Franco, who bestowed on him the equivalent of a Spanish knighthood, which came with a huge red presidential sash. That particular vestment became my cousin's proudest possession, and there would be no occasion too slight for him to wear this vermilion keepsake.

He stopped off in New York one winter to see us in Forest Hills—an event equivalent to King George visiting at Versailles. New York City had just endured a week-long torrential downpour and devastating flood conditions beyond the imagination or experience of any Spaniard or California native. My last and fondest memory of cousin Buddy was his grand emergence from a limousine, immediately followed by him and his resplendent sash disappearing in the swamp of deep mud that had been our driveway the previous week.

II

How my interest in music developed is still a mystery, but, while in Forest Hills, I became aware that a famous band leader, Hal Kemp, lived about three doors down from us. There was also a nightclub named the Ideal Spot located in the "raunchy" outskirts that overlooked the entire town. The pianist Art Hodes had a trio there with Rod Kless on clarinet and bassist Jimmy Butts. Dan Priest, who was my best friend, and I would go to this club to hear Hodes, and it was a seductively exciting, tantalizing experience for us youngsters to be in a nightclub, with musicians and older people, in a "forbidden" section of the city.

When I was thirteen or fourteen, and had a birthday, my father had no idea what kind of gift he should get me. He spoke to a fellow named Jimmy McGill, who was a local dance band leader about twenty years older than I was at the time. He suggested my gift should be a phonograph player and a few records, which my father, much to his subsequent dismay, then got for me. The records included Tommy Dorsey's "Song of India," "Sometimes I'm Happy" by Benny Goodman, such Hal Kemp current hits as "Lamplight" and "Got a Date with an Angel," and an album of 78s entitled *Rippling Rhythm Time*, which included a few of the Bluebird singles by Shep Fields.

I began to listen, first almost indifferently, and then with increasing obsession until I was completely hooked! Somehow I realized these were not corny dance bands but substantial musicians creating something very important. I began to hear some Fats Waller records on Victor such as "It's a Sin To Tell a Lie," and those really turned me into a jazz fan. I heard Frankie Trumbauer with the Paul Whiteman Orchestra, and through those recordings, I learned of Bix Beiderbecke. At this time there were about four or five musicians I listened to—and I just wanted to hear every single note they recorded.

From then on, when I would come home from school, I went right to my room and listened to records until dinner time. My mother was positive her son had developed an incurable mental condition, and my father, who wasn't at all sure what was happening, could only say, "He likes music, so let him listen to music." Over those few years, I probably heard every jazz record ever made, and in my mind I can practically hear all of those records and improvised solos today, and still know them intimately.

I became seriously engrossed in listening to records, especially jazz, and to live music with my by-now equally possessed chum, Dan Priest. We started to go to Nick's, a club in Greenwich Village that featured small Dixieland jazz groups with Muggsy Spanier, Bud Freeman, Pee Wee Russell, Bobby Hackett, Wild Bill Davidson, etc. Any "name" you heard of was there, and it was quite an introduction to live jazz for two enthralled pubescents.

Although we were still too young to socialize with our new heroes, we would nonetheless arrive at Nick's at around 9 p.m. (very early in those days) just about every night, get our front-row seats, order Coca-Cola, and stay listening to the music until 4 a.m. And we'd go up to Harlem to see Chick Webb at the Savoy, then hang out at jam sessions afterwards.

There was a place right across the street from Nick's called Julius's (which is now, like so many other former musical landmarks in the Village, a gay bar), where the musicians would go to relax, drink, and fraternize between sets. That was where I became close to many of the great musicians of the era.

Among the episodes I remember from this period was one that involved a man named Dan Qualey (who was as far right musically as the senator and vice president with almost the same name was politically decades later). Passionately enraptured with the style known as Boogie Woogie, Qualey owned a record company called SoloArt that specialized exclusively with the piano giants of the idiom such as Meade Lux Lewis, Albert Ammons, Pete Johnson, and even Jimmy Yancy.

One night at Nick's, we were all at the bar about to go to a party given by the young jazz critics George Simon and Leonard Feather, who were then sharing a Greenwich Village apartment. Dan Qualey, who was not invited to this gathering, overheard our preparations to leave and, at a discreet distance, followed us to the Simon-Feather abode. When the door opened, and we were warmly greeted by George Simon as invited friends, Dan Qualey suddenly materialized behind us, rushed in, and instantaneously punched Leonard Feather in the nose!

We called an ambulance that took a quite-bloodied Leonard Feather to St. Vincent's Hospital nearby. Soon after, we learned that Qualey was enraged with Leonard because he had dared to criticize his beloved Boogie Woogie recordings, and eager for an opportunity—which we had unknowingly provided—to wreak righteous vengeance on this musical infidel.

During the next couple of years, Leonard was unalterably convinced I had instigated this incident for the amusement of our circle. Since then we have remained very good friends, but at the time it was a long while before he could be persuaded of my complete innocence in the matter.

This was typical of a very strange era in jazz where every musician and fan was vehemently loyal to his personal preference. If Dixieland was your thing, Swing was anathema. Boogie and Be-Bop, impossible. Everyone had their own inviolate, favorite category, musicians, and recordings, and nothing else could co-exist with these personal icons. It was like an endless brawl between fanatics who, if they could see how nonsensical they looked, would have found they had much more to share with each other, and could have helped this cultural outcast called jazz music find its eventual wide popularity a lot sooner.

III

Another night Dan Priest and I were in Nick's sitting at the bar next to a rather attractive blonde-haired lady who, we soon learned, was Wild Bill Davidson's wife. We were having a most innocent, friendly and very respectful conversation with the wife of one of our idols when the band finished their set and Wild Bill angrily stormed toward us. I instantly feared he was about to hit me, but he punched his wife instead—knocked her right off her chair—and screamed, "I don't want you talking to these guys!" He was to be sure a daring trumpeter, who would take many liberties with the harmonic and rhythmic norms of the era, but this was proof that Wild Bill earned his name more by his social excesses than by his adventurous musicianship.

The Village was phenomenal as a center of around-the-clock activity, and for days on end I could never (nor even want to) leave, with so much classic jazz as well as many other kinds of great music available at the numerous clubs you could walk into just about any time of the day and night. I had the best possible introduction to jazz in Greenwich Village when it seemed that those few square miles were the true epicenter of musical energy in the world.

Once in the late 1930s, when in the Village alone I met Monte Kay, who years later married Diahann Carroll and managed the Modern Jazz Quartet and comedian Flip Wilson. Monte and I used to hang out at a couple of places together, and one morning, as the sun was rising, we ran into Huddie "Leadbelly" Ledbetter, the legendary blues singer and composer. We all sat down on the curb at I believe Seventh Avenue and 4th Street, and Leadbelly sang the blues for us with his guitar, himself in the middle and Monte and I on either side of him. There were the three of us, sitting on a curb in the Village at dawn, and Leadbelly singing his ass off. I guess this was the time he had just been released from prison through the efforts of Alan Lomax, the also legendary blues fan and anthologist.

There was always a lot of bouncing around in Greenwich Village during that period. I remember attending a jam session at the soon-to-be-very-famous Village Vanguard with Max Gordon, the eternally juvenescent, cherubic owner of the club. The only musician I recall at this occasion was the bassist John Simmons, whose daughter Sue is the perennially popular television news-anchorperson on NBC Channel 4 in New York City. These jams were run by Harry Lim, a diminutive gentleman from Java who soon after founded the invaluable Keynote record label.

Somehow a fight broke out during the music, and tables and chairs were literally and liberally flying everywhere inside the Vanguard, unquestionably one of the smallest nightclubs ever known. During the melee, John Simmons actually picked up all 4 feet and 6 inches of Harry Lim and threw him entire length of the club. Dan Priest and I were by this time experienced enough to manage our way out of the club and up its extremely narrow stairway that not uncommon (for that time) evening in the Village.

Dan and I were also frequent visitors at the fabled Village nightspot Cafe Society, which was owned by Barney Josephson, who always had a cachet of subversion about him. (I think he was one of the first organizers of the Communist Party in New York City.) He was also a good friend of John Hammond, the hobbyist who discovered and recorded such immortals as Count Basie, Benny Goodman, Billie Holiday, Lester Young, and Charlie Christian, and decades later, Aretha Franklin, Bob Dylan, and Bruce Springsteen.

At Cafe Society, we heard Billie Holiday, the Golden Gate Quartet, Teddy Wilson, and perhaps the most interesting group to ever develop out of that nightlife, The Revuers, with Judy Holliday, her friends Betty Comden and Adolf Green, and their friend, a young pianist and student composer by the name of Leonard Bernstein.

During the same time, I would spend hours at my parents' home in Forest Hills on the third floor of the house, which I had all to myself. Dan Priest and I would play my rapidly burgeoning collection of 78 rpm jazz records loudly and endlessly. From the moment I came home at three o'clock from school, until dinner time three hours later, I was in my room listening uninterruptedly to Jelly Roll Morton, Johnny Dodds, Benny Goodman, Bix Beiderbecke (you name it, I had the record). No one could figure out this kid who was doggedly buying and listening to jazz records when he should have been playing baseball like every other boy his age.

Our neighbor in this still very posh area of New York City was Chester Conn, a major music publisher at that time, and he was as confused as my father and mother with my constant interest in and manic listening to jazz records. My father would always run into Chester Conn and say, "What is it with my kid?" And Chester could only advise, "Leave him alone. He loves music. I know a lot of people like this, and I can never explain it." No one ever really knows how and why anything becomes the defining aspect of a person's life but, in my case, it certainly happened with music. Benny Goodman played clarinet, Duke Ellington played piano, and Bob Thiele played 78s.

Throughout the 1930s, this was the pattern of my life. I also spent vast amounts of my free time and allowance buying records and in music stores looking for rare items. In those years companies did not reissue important recordings as they constantly do now, and I was forced to search long, hard, and persistently for treasures by Benny Goodman, Bix Beiderbecke, Fletcher Henderson, Ellington, etc. A lot of time was spent at the Commodore Record Shop on 42nd Street, owned by Milt Gabler in those days before he went to work for Decca Records as a jazz producer.

Another habitual stopoff was Steve Smith's Hot Record Society record store on Seventh Avenue at 53rd or 54th Street. It was there around 1937, that I met Jelly Roll Morton (of all people!) in a listening booth—in that era record stores had booths where you could listen in private to recordings that interested you to decide whether they were worthy of purchase. My meager recollections of that meeting include a gold-toothed Jelly Roll insisting that he alone invented jazz. I had met the inventor of jazz. Wow.

The Benny Goodman records had another affect. As a teenager I learned to play the clarinet and soon had a 14-piece band that played dance jobs at schools and beach clubs throughout the New York City area. We exceeded in adolescent enthusiasm what we conspicuously lacked in musicianship, and anyone who heard us would agree we had one attribute that separated us from all our contemporaries and competition: we were the LOUDEST band on the East Coast, or any continent for that matter.

Evidently, to hear us was to hate us since we never had return engagements, and the one time I used my familial "clout" to wangle a job at Lundy's (a cavernous restaurant renowned not only for its food and colorful waiters, but for its seismically reverberant sound level), the band was fired midway in our first set because so many customers had started to complain about how loud *we* were!

Another time I booked the band at the very exclusive Atlantic Beach Club on Long Island, and I decided, as this was an authentic "big time" gig, we had to have a more professional look. I brought some long wooden boards and put them on gantries so the brass players could sit elevated above the saxophones—just like a real band.

All evening my trumpet player, uncontrollably possessed and impressed with where we were working, kept shouting down at me between numbers, "Let me sing one! Let me sing one!" Finally, on "Music, Maestro, Please" (and mainly to shut him up), I motioned to him that his big chance had arrived and to come down front for the vocal. In his eager haste, he knocked over the plank he was sitting on, and the entire brass section crashed to the floor in a loud confusion of bleats, blatts, breakage, and instruments flying. Neither Spike Jones or Mack Sennett could have done it any better.

I also ran Sunday afternoon jam sessions at a 52nd Street club called Kelly's Stable and consistently lost money. But I kept it going for quite a few weeks, and great jazz players like Coleman Hawkins and Lester Young—all the fine musicians who were playing around the city at the time—would come by and pick up their $30 for an afternoon of unrestrained jazz, free from band leaders and written arrangements. The jam sessions were a financial fiasco, but a lot of good music was played and heard.

One of my most treasured memories is of a piano player by the name of Shorty Nadine who showed up one afternoon. He swung so hard and was so inventive, he became a regular. The major jazz musicians would wait around until he arrived so they could play with him. About the only thing he seemingly couldn't do was sing, but we all learned of this addditional skill a short time later when he changed his name to Nat Cole.

Dan Priest and I were constantly listening to jazz and discovering new (to us) musicians. Around 1939, we decided to publish a magazine (the very first in its category) which we called (what else?) *Jazz*, having secured an ancient linotype machine.

The first issue was really interesting as we were just two teenaged jazz fanatics who, in the zealous belief nothing was impossible, somehow obtained enough articles and pictures to fill up a magazine, and persuaded the finest writers and photographers in the business to contribute their efforts. They weren't interested in money (not that we had any to pay them) as much as they wanted their work published in some fashion. Some of these names might not mean as much now, but we had such esteemed writers as Charles Edward Smith, Charles Paine Rogers, Fred Ramsey, Jim Weaver, and Jake Trussell, Jr.

That first issue was also notable for a detailed article on Scott Joplin, whom Dan and I knew little about to that point other than he had written a couple of hit song "rags" many years before. We managed to find a rare picture of him and inadvertently ours was the first publication to recognize Scott Joplin as a historical jazz figure. Over the next few years *Jazz* actually grew to the point where it became an important magazine with 8-by-11-inch glossy covers and inside pages.

I also went frequently to Harlem, where I spent almost as much time as I did in the Village downtown. During that period anyone could comfortably visit that neighborhood, and I would go to the famed Savoy Ballroom, get in with no trouble, and move up to the bandstand to be immobile for hours listening to the orchestras of Chick Webb, Andy Kirk, and Jimmie Lunceford. It was an exciting time and a beautiful feeling to be there with the mixture of blacks and whites and never any problems or tensions, just great music.

Another attraction in the vicinity was the historical Harlem hot haunt called Minton's Playhouse, where I listened to Thelonious Monk, Dizzy Gillespie, Joe Guy, Kenny Clarke, and those two immortal Charlies, Parker and Christian, while they and their collaborators redefined forever the vocabulary of jazz every night.

3   Airplay

A T age fourteen, I promoted myself into my own radio show, a 15-minute program once a week on WWRL, then a foreign-language station in Woodside near my Forest Hills neighborhood. I didn't get paid but was allowed to play jazz records for a quarter of an hour. The show was named "The Jazz Band Ball," and must have gotten wide "coverage" as I remember receiving telephone calls from the legendary Cleveland radio personality Bill Randle, who was incredulous that any radio station would permit recordings by such black jazz musicians as Jelly Roll Morton, Johnny Dodds, Louis, Duke, etc. to be played on the air.

The show remained on the station for some time, and I remember what is now a funny experience. I was still, after all, a kid living at home. Upon my return from WWRL one afternoon, there was a summons on the front doorknob of my parents' house. Glenn Miller was suing me for libel because of an opinion I shared on the air about the band. I had compared it with the superlative swing machines of the day, such as the Fletcher Henderson Orchestra and Count Basie, and said the Miller band did not compare with the excitement and beat these ensembles provided.

I panicked, big-time. I was this teenaged kid being sued by

one of the biggest stars in America for a million dollars. In blind terror I immediately rushed downtown and showed the summons to my father's lawyer, who read the first few words and burst out laughing. "This is a phony. Somebody just bought a summons form in a stationery store and filled it out."

Later, I discovered the perpetrator to be none other than my best friend Dan Priest, who decided to play this immense joke on me and added a few years on to my life right then, and probably took a couple off later on. Some joke. Some friend.

I eventually wound up on another radio station, WBYN, a Brooklyn station with studios in midtown Manhattan on 44th Street, where I played and talked about jazz records for an hour, five nights a week between midnight and 1 a.m. I really enjoyed that. A whole hour every night to really stretch out.

It was during this sojourn that I first met the Ertegun brothers, Ahmed and Neshui, who eventually built an empire named Atlantic Records. They just wanted to congratulate me on this unique occurrence—a one-hour radio show devoted to jazz records. The Erteguns appeared at 11 p.m. one night and found my engineer, who was in another part of the building from where the broadcast studio was located, and asked to meet me.

Whether the coffee—the traditional sustenance of all radio engineers—had taken effect too little or too much, all the engineer could see was two Arab gentlemen with thick accents and unpronounceable names in expensive business suits. He immediately assumed, in those days when prewar intrigue was probably as prevalent as it was imagined throughout the city, that the station was now the center of some new subversive media conspiracy to be hatched on an unsuspecting population of insomniacs. Between records I convinced my engineer that no secrets from locked-up radio offices would be violated if he would escort these charming young Turkish aristocrats—whose father happened to be their nation's ambassador to the United States at the time—to where the studio was located.

My next show was on WHN, a powerhouse (50,000 watts) radio station, where my hour of jazz was preceded by a hot disc jockey named Dick Gilbert who played pop records, and followed by Alan Courtney, another "torrid platter turner." (Actually, this was many years before on-air personalities were called "disc jockeys." In those days, we were "announcers.") I was getting more

mail than all the other jocks at the station combined. It seems there
were more jazz fans wherever the signal reached than anyone had
imagined, and even to this day the music business continues to
minimize the audience for jazz: among the most loyal and supportive in any category.

The show lasted a year until the pressures from the other
announcers—all of whom had union membership, contracts, se-
niority, and the paranoia the first three inexorably encourage—
forced the management to take me off the air. This was to be my
final fling at a radio career for I soon exchanged the politics of one
music medium for another.

4 | Off Base

WHEN the naval base at Pearl Harbor was attacked, President Roosevelt had only one choice, and directly, many thousands of young men like myself had one of two: either be drafted and most likely shipped someplace nobody ever heard of where people would very seriously try to kill you; or enlist and perhaps be sent somewhere less inhospitable.

To this day I continue to be amazed at my good fortune. During that period I was an incredibly lucky enlistee who somehow ended up assigned to the Manhattan Beach Coast Guard barracks in Brooklyn on New York harbor near my home and beloved Manhattan nightlife. Not that I landed in a country club setting by any means; without question I was in the real army: an army with two-tier bunks in shacks with no heat or hot water, and nasty military neanderthaloids who woke everyone up at 6 a.m. by clanging metal batons on the brass frames of the beds to move us toward our main activities of each day: freezing and standing at attention for hours and hours.

As the base was surrounded by water, it was always freezing. Illness was a constant condition, and I even remember fellow Guardsmen actually dying from the omnipresent cold climate we were all exposed to. Everyone had a 101° temperature, and since this was the army, you had to have at least 102° to get into the infirmary! Still, most days I kept in mind that this was preferable to dodging bullets.

The closest I came to warfare were my run-ins with (and run-aways from) the Military Police known as the SPs (Shore Patrol) who, in the war years, were as ubiquitous in the Metropolitan Area as any other uniformed service. I certainly kept enough of them busy a lot of the time all by myself, since the army, in its wisdom, had granted me a privilege known as "free gangway" as a result of my proximity to my home, and my eventual appointment to the "Military Morale Department" on the base.

"Free gangway" meant I could leave the base in uniform after 5 p.m. as long as I returned no later than 6 a.m. the following day, a perk I shamelessly used and abused without hesitation. As a "morale sailor," I was billeted with the musicians in the Coast Guard band, and we would go into the city to hang out at jazz clubs whenever I wasn't using the time to make records for my Signature label, which I had begun before I enlisted.

The band at the base had many fine musicians, such as Lew Brown, who subsequently became Jerry Lewis's longtime pianist/ arranger/conductor, and Shelly Manne, a match-thin bundle of perpetual enthusiasm and energy who would soon become one of the most famous jazz drummers. (In fact, Shelly made his first records and reputation during our frequent "night patrols" into Manhattan when he was the drummer on sessions I produced with Coleman Hawkins, Lester Young, and Eddie Heywood, among many others.)

One Guardsman who would wander around Manhattan with us was Roy Sedley, a young comic who had worked at the notorious Club 18 on 52nd Street with another comedy newcomer named Jackie Gleason. Since Gleason loved jazz and was around the musicians as much as myself, he was another eager companion on our nightly expeditions throughout the city. Roy Sedley was also the first husband of Harriet Hilliard, a popular band singer who became Ozzie's Harriet. Her earliest spouse, who was one of the funniest people I ever knew, is only remembered now as a trivia answer.

Of course, since I was as spoiled and arrogant as anyone

would be who came from a well-to-do restaurant family in Forest Hills, had friends like Duke Ellington and Billie Holiday, and couldn't wait to leave the Coast Guard, I was always in trouble. I would get falling-down drunk and/or lose my sailor's cap (an even graver military violation as that constituted being out of uniform) and made the Shore Patrol's "Top Ten" list. They would try to arrest me stumbling out of clubs except when I retained enough of my faculties to outrun them. Either way, we all got to know each other very well.

I recall one time when they were (as usual) too fast for me. That night I was incarcerated at the municipal jail cells on 52nd Street "reserved for Armed Forces Personnel" before my transfer early the next morning to Pier 91 on the West Side of Manhattan where all the "military offenders" of the previous night would be corralled pending any further disciplinary action.

Along with 2000 others, and surrounded by armed guards every foot of the way, I was marched out for breakfast next to an all-American, clean-shaven, blond and blue-eyed crew-cut youngster who seemed as unnerved as I was by all the armament pointed at us. When I asked my fellow prisoner why there were so many guns, he whispered that there were extremely vicious criminals among us who would do anything to escape. During our meal, he asked what I was in for, and after I told him I was being detained on drunk charges, my new buddy suddenly became wild-eyed and started to boast he had stolen a jeep in Iran, which he used to run down and kill five members of the local citizenry because (now shouting) "I hate them A-rabs!" Terrific, I am having breakfast with a mass murderer on a Hudson River pier. (Later I learned he was sent to Leavenworth Penitentiary for 25 years.)

With my Military Morale Department assignment, I defended my nation with a variety of weighty responsibilities. I was, for example, in charge of the Friday night dances at the base where the Coast Guard band would play for the happy couples. Specifically, it was my task to choose which ladies, from over 7000 Metropolitan Area applicants who had mailed in their pictures, ages, names, and addresses, would be lucky enough to be admitted onto the Manhattan Beach perimeter for an evening of dancing with several thousand female-deprived sailors.

I approached this solemn duty with my usual diligence. Several of my chums and I would sit around for hours in a lackadaisical stupor flipping photos like coins and saying, "This girl's O.K., this one's not. This bow-wow belongs in a kennel," etc.

My stewardship of the base canteen was approached with equal resolve. After every lunchtime, three sailors under my "command" and I would stand behind a giant candy counter as all 10,000 guardsman on the base became simultaneously fixated with a need for sweets.

We had an inexhaustible supply of Hershey and Mars bars, Snickers, etc.—everything the civilian population we were protecting could not get. Thus, it was the four of us against a rampageous daily horde of sugar-crazed paladins.

The lines were so long that, whenever a sailor actually got to the promised land of our counter, he would order bulk quantities ("I want 5 of . . . , can I get 10 of . . . , 20 of . . . ," etc.), and in the continually frenzied scramble, my staff, inculcated with my inspiring leadership, joined me in throwing handfuls of candy over the table and, without ever counting, putting whatever money we were handed—usually a couple of dollar bills—into the cash register.

Everyone would pay $2 and leave with $10 worth of candy, and we didn't give a shit. At the end of each month, a different Chief Canteen Petty Officer would arrive to inform us, "You guys are short $100,000, what happened?"—a query we found invariably hilarious.

Thus, as a result of such personal heroism in those darkest times of global hostilities, the world was once again made safe for democracy, *Howdy Doody*, the Edsel, and jazz records.

5 My Signature

I

MY very first label, Signature Records, started when I was seventeen and in my last year (1938–39) of high school. I began to hear so many musicians who weren't being recorded, and me on my white horse was determined to document their stuff. It was a sort of perseverance and aberrant dedication to collect and make jazz records. Everybody was pressing records, all kinds of music. I was always impressed and envious there were small labels around that made jazz records. They probably weren't selling too many of them, but at least they were recording jazz.

Actually, the year before, I impetuously and enthusiastically booked pianist Joe Sullivan into a studio for what would have been my first production before I learned that this required more money than I imagined or had. Embarrassed to ask my parents for the requisite funds or admit to Joe Sullivan I couldn't pay him, I never showed up at the session. It was many months later that I was finally able to make my debut as a record producer: four sides with my friendly neighborhood jazz trio (from the Ideal Spot nightclub in Forest Hills), the Art Hodes threesome.

I was taking an art class in my senior year and, as I was interested in recording, I designed record labels for two imagined companies. One was (honest!) Capitol Records with a dome for its logo, and the other was Signature with a musical treble clef as its symbol. I tossed a coin and it came out Signature, where I then put out my recently recorded Hodes sides. This was years ahead of Capitol Records, Mercury, and the other record business giants.

All I knew was that you hire a studio and get an engineer, who then puts up a few microphones, and that was it. There was no tape then; the music went from the instruments into the microphones and onto lacquer discs. Through youthful and energetic diligence, I located the Scranton Record Company in Pennsylvania. It was originally a button factory that had since become knowledgeable in the still-arcane process of pressing phonograph records. The man in charge of the company was named Roy Marquadt. Through the years, he gained a reputation as one of the pioneers of the record industry.

Scranton would press records for anybody, and after they pressed up my 78s, they would ship them to my father's house in Forest Hills. As I was (out of necessity) my own distributor, I would then hit the record stores in Manhattan on Broadway and on Madison Avenue and actually sell my shellacs out of the back of my father's car directly to the dealers.

Advertising was a couple of small ads in music magazines, and I would have one or two record store owners in each of the major cities. At that time, those dealers would buy perhaps two copies of almost every record that came out, especially for their collector customers. When my pressings arrived from Pennsylvania, I would take them and my orders into the top-floor bathroom, wrap the records into cartons, and run the gum tape through the sink. Then I would mail out the records at my neighborhood post office, and bill the customer.

For my local accounts, I'd deliver the records personally and get paid for them. I sold to every store in New York City. In those days, there was Schirmer, the Commodore Record Shop, an outfit on Madison Avenue called Marconi's, and at least fifteen record outlets along Broadway from 42nd to 57th Street. In fact, at one of the Broadway stores where I had just sold ten records to the owner, he called out to another kid my age in the shipping room and said, "Hey, Frank, go out and help this guy unload these records out of his trunk." That was when I first met Frank Military, who would become one of the most powerful music publishers in the business, and the publisher/rescuer of "What a Wonderful World" years later.

II

I did record dates with everyone I could. One of the first small combos I recorded I called the "Chicago Rhythm Kings" only because I remembered a long-ago group by that name that had included Bud Freeman, Gene Krupa, Joe Sullivan, and Frank Teschemacher and recorded for Okeh, one of the very earliest labels. The drummer of my "Rhythm Kings" was Chicago legend George Wettling, who didn't show up for the session, but trumpeter Marty Marsala, Rod Kless on clarinet, Art Hodes, a guitarist, and a bass player from Chicago, Earl Murphy, did.

The first big band sides I did were with Yank Lawson on trumpet and simply called "Yank Lawson and His Orchestra," with arrangements by Bob Haggart and Matty Matlock. The tunes were "Sensation Rag" and "Sugar" in the Bob Crosby Dixieland style that was then the rage.

In addition to Yank, I started to record with Pee Wee Russell, Eddie Condon and, through a geographical coincidence, James P. Johnson. I had first become aware of him through the Scott Joplin feature story in our *Jazz* magazine, which extolled Johnson and Willie "The Lion" Smith as his artistic equals. As a result of my callow curiosity, I listened to a few James P. Johnson records, and then discovered that he lived near my Forest Hills home. Brazenly, I just showed up at his house one day and knocked. A little gray-haired lady opened the door to hear me announce: "You don't know me, I'm Bob Thiele, and I just wanted to meet the great James P. Johnson!" She invited me in, and soon the hulking presence of the great James P. himself came down the stairs. I don't want to say he was ugly, but with his gargantuan head, he certainly wasn't handsome. He was, however, charmed by my fervent enthusiasm and treated me to a one-hour concert on his living-room piano, playing all his compositions. We became immediate friends, and I recorded him with musicians such as Yank and Pee

Wee, as well as on some of his most famous and influential solo sides.

Although I was the youngest record producer any of the musicians had ever seen, I was always well received by them. My personality was such that they knew I was for real and not just some jerky kid. I was really into the music, and I think they respected that. I've always gotten along with musicians very well. That's the reason I've been able to work with so many of them and maintain those relationships.

For Signature, I recorded the classic Coleman Hawkins ballad rendition of "The Man I Love," and the sides with trombonist Dickie Wells and his wondrous saxophone sidekick from the Count Basie Orchestra, Lester Young. We usually had to record these musicians at two or three in the morning, after the jazz clubs where they were working closed, and during the only time I (and Shelly Manne) was not obligated to protect our country.

I'll never forget "The Man I Love" session. At the time most companies would press only ten-inch 78s. But the reason this "The Man I Love" turned out as one of the few twelve-inch 78s in commercial release was simply because Coleman Hawkins wouldn't quit. He'd take extra choruses. We were recording at radio station WOR on Broadway in the middle of the night, and right during the middle of that record a cleaning woman walked in with a mop, intent on cleaning the studio. I literally walked out into the studio, put my fingers to my lips to be quiet, then held her arms. They played, and I'm holding a struggling cleaning woman while one of the most immortal solos in jazz history was being recorded. That's a part of "The Man I Love" not too many people know about.

III

While I was in the service my father, who had moved his chocolate sales business to an office in downtown Manhattan at 601 West 26th Street, offered to help me out with what he still considered to be my "hobby." He told me, "Look, Bob, I'll try to keep this music business thing of yours going. If we get any orders for records, we'll ship them out, and I'll order more as we need them." What then happened was that "The Man I Love," which I had released as a twelve-inch 78 record at the going retail price of $1.50, really took off.

My father's office was listed as the address for Signature Records, and all of a sudden, record dealers from everywhere—Harlem, New Jersey, New England, and beyond—would show up in person wanting to buy *hundreds* of copies of "The Man I Love." My dad started to get excited and began entertaining thoughts that his crazy kid had fallen into something very substantial. Record dealers—*cash customers!*—were constantly at his door, and he was selling more records than chocolate and continually phoning to Scranton for more stock.

Just as quickly, things started to fizzle out. Johnny Mercer, band leader Stan Kenton, and Glenn Wallichs, who owned the largest record store in Hollywood, formed Capitol Records and, because a war was on and they could not obtain pressing materials or machinery for themselves, bought the Scranton Record Company and threw out all the independent labels. (Only a few years later, Scranton became the chief manufacturing center for the Capitol Records mega-conglomerate.) Somehow, with our rapidly depleted inventory, we barely squeaked through and survived to welcome the postwar music explosion with which the United States, deliriously happy in victory, celebrated its economic and moral predominance in the new world forged with its forces and allies.

Many people wanted to invest serious money in this upstart record label that had survived both another war to end all wars and the war to end all independent record companies. And so Signature Records went public, and we decided to build our own factory to press our records.

We started at 26th Street, in a giant warehouse named the Starret-Lehigh Building, where we would locate those most necessary 78 rpm record presses. I asked Steve Sholes, a friend of mine at RCA Records who later had the distinction (and eternal job security) to be the executive who signed Elvis Presley to that company, where I could find anyone who knew how to build and run a record press. He knew a most qualified gentleman at the RCA factory in Camden, New Jersey, named Ray Wickes. He suggested I tell him that Mr. Sholes from RCA asked that we meet, and, if he was agreeable to be hired, that he would be the perfect man to supervise the construction and operation of our presses.

Soon after we hired Ray Wickes. Unfortunately, it was not Mr.

Wickes *Senior*, but his son, who, in a farcical-mistaken identity scenario the Marx Brothers would have envied, joyfully accepted our offer of employment and, in his glee to be liberated from Camden to the New York City bigtime, neglected to inform us that he knew nothing about building a record plant.

Under the supervision of our highly recommended "expert," we installed plumbing, bought the correct dies, obtained our supply of shellac "biscuits," and somehow did everything right (we thought). At the same time, we met a man named John Busey, who was the president of the General Electric Supply Corporation, which had 144 distribution points in the country, and wanted to own the largest distribution network in the history of the record business.

The combination of our aggressive eagerness to be a major industry player and the desire of GE to replace RCA as a record distribution giant was an instantly attractive, heaven-made merger of two companies with the "perfect fit" of assets, delivery systems, and goals. We would make and press Signature records, and GE would distribute our product and set the new industry standards we could so clearly envision in our imminent, aligned future. (Decades later, in the merger-frenzied 1980s, GE would actually buy RCA Records, along with the NBC radio and television networks, and the other various RCA manufacturing and communications businesses, only to promptly sell off RCA Records to BMG, a German mega-conglomerate even larger than GE was at the time.)

To ratify our industrial marriage, we arranged to manufacture our first Signature 78s to be distributed by GE in full view of the national press at a lavish party in our 26th Street factory.

Before this extraordinary saga continues, a few words to describe the actual record pressing process would be appropriate. A wet shellac "biscuit" would be inserted between two tremendous dies or stampers, one on top and the other below. Steam would then flow through the machinery to flatten and dry out the shellac, and when the dies/stampers were pulled back, an impermeable 78 rpm record was the perfectly formed issue.

With reporters and photographers anxiously present to witness this commercial miracle, the younger (and now completely overwhelmed) Ray Wickes "froze," paralyzed with fear and confusion and the sure knowledge that his fraudulent credentials were about to be exposed to the entire non-Communist world. I, brazenly confident that no barrier existed to deny me my predestined Olympian success, said, "What's the big deal, I'll press the damn record myself."

The assembled press then saw the fastest rising executive in the record industry immediately cover himself with melted shellac. A head-to-toe monument to hubris.

The song did not end, but neither did the comedy and tragedy. Through various connections we located a George Jaycox, who worked at the Bridgeport, Connecticut, Columbia Records plant, and decided we should hire him to organize the zany zoo that was our factory.

One midnight, after he was able to leave his Bridgeport quarters unsuspected, Mr. Jaycox arrived at our building, looked at our equipment and said, "Very nice, but I hope you're not trying to press any records here." "I already did," I defended. "Well, you're lucky to be alive, Mr. Thiele. Were you aware that you installed household plumbing to run the necessary 300 plus degrees of steam for pressing? You could have blown up the entire building."

George Jaycox did come to work for us. He soon organized our plant so well that we began to think about expanding our capacity, and perhaps lease our services to other companies because GE was ordering our records in quantities beyond our wildest fantasies. What we were not aware of was that GE's chief priority was household appliances, and these 144 distributors when offered records at a (even then) cheap 45 cents a single unit from their corporate home base, would each loyally order a thousand Signature 78s, store them in a basement, and go back to selling products they understood, such as toasters and refrigerators, to their customers.

The envy of the record industry, we were moving tens of thousands of our jazz records into a state-of-the-art distribution network above anything the music business had ever seen. We also had expanded into pop music with singles by Johnny Long, Skinnay Ennis, and Alan Dale. What we did not realize, of course, was that we were just *shipping* records, not selling them, and rapidly depleting our financial resources to manufacture these remarkable quantities of records that were never to be seen by any potential purchaser. (Years later, the industry also learned this painful difference between shipment and sales. The Recording Industries Association of America (RIAA) would certify a gold record as a half a million units *shipped*, and a million units qualified as platinum. Inevitably and promptly, this became a bonanza for counterfeiters. The joke became, "We shipped gold, and got returned platinum." Now, gold and platinum certification is based only on actual sales, thank you very much.)

In any business, certain words raise flags of panic, and in ours none more so than "returns." All of a sudden, we were deluged with returns from our distributors, who now needed their basement storage space for the newest GE freezers, radio consoles, etc. We realized, too late to rectify certain calamity, that we had gone through our entire financing and had made records that never reached customers.

IV

My career as a record executive had almost ended before it began, but it was an invaluable lesson to learn the essential distinction between shipping and selling. Some additional and indispensable advice also came my way during this experience.

During our frenetic period of pressing and shipping vast quantities of records to satisfy our distribution orders, we did indeed start to provide record pressing services to other independent labels. This "side business" soon became sufficiently profitable to persuade us to build another pressing facility in Shelton, Connecticut.

One gentleman who became an account for us was Eli Oberstein, a veteran industry eminence who, as an artists and repertoire director for RCA, had been responsible for the recordings made by Glenn Miller, Tommy Dorsey, and Benny Goodman in their prime periods of success. He then left RCA to form his own label named Royal and, after he set up a warehouse/office in Connecticut near our second plant, asked us to press his records.

With our proviso that all transactions be C.O.D.—it was, and still is, standard industry practice to treat all independent labels as C.O.D. customers—we entered into a comfortable relationship that was simplicity itself: we would press 10,000 copies at our Connecticut plant, ship them to the Royal Records warehouse in the same state, and be paid in cash upon delivery. This continued until the last shipment when we did not get paid: Royal went out of business, and, after unsuccessful attempts to resolve the open balance, we ended up in a lawsuit. (I don't even remember if we ever collected any of our money from the action.)

Sometime later, on a train to New York one Sunday, I met Eli. I know it was a Sunday as Eli told me it was the only day he could come into the city because his wife was suing him for alimony arrears, and blue laws prevented his being served on Sunday.

After I commiserated with him about his domestic troubles, I could not resist putting it to him point blank. "Eli, why didn't you pay us? We shipped you a load of records, which you then sold, and you never paid for them." He replied, "Bob, you gotta learn the record business. When you know you are not going to do business with someone anymore, don't pay the last bill!" While I never put this "advice" into practice, I never forgot it either.

V

When Signature Records was prospering and I was approaching a zenith of prowess and power as a hit-making star executive, I literally kidnapped a major industry figure for several weeks, transporting him over many state lines. Fortunately, the "victim," my friend and prominent music business magnate Frank Military, chose not to reveal my criminal act to the authorities or otherwise press charges.

Ever since we were both teenagers who knew very little about the music industry other than we had to be in it, Frank and I have remained close friends and professional comrades. When we first met, Frank worked at the shipping room and loading dock in one of the many big record stores that dotted Broadway between Times Square and Columbus Circle, where I, with my first record label, would sell my Signature Records jazz shellacs out of the back of my parent's car.

Currently, five decades after, Frank Military is a senior vice president of Warner/Chappell Music, the eminent and immense music publishing conglomerate. Respected as among the most effective, influential, and creative music professionals in the industry, he has enjoyed close personal involvements with just about all the important American popular song vocalists, including Bing Crosby, Barbra Streisand, Perry Como, Johnny Mathis, Peggy Lee,

Bette Midler, Tony Bennett, and his dear friends Frank Sinatra and
Dean Martin. (At the earliest beginnings of their careers, Frank,
along with Dean Martin and Dean's manager, lived together in one
room at a squalid midtown Manhattan hotel where they would
take turns sleeping in the bathtub while the other two shared
the single bed.) Frank has also been significantly associated with
virtually every major theater and film composer and lyricist such
as Jule Styne, Sammy Cahn, Stephen Sondheim, Cy Coleman, Alan
Jay Lerner, Johnny Mercer, Harry Warren, Jimmy Van Heusen,
and Cole Porter, among major songwriters.

With the mutual realization that both of us bright-eyed youngsters would survive and flourish in the music industry, Frank and I
rapidly became very close friends. Soon after, he was a second son
to the Thiele family: an always welcome guest at our Atlantic
Beach cabana, going with us to the tennis matches in Forest Hills,
taking pies home from Lundy's, and even traveling with my father
Herbert to Scranton, Pennsylvania, to help carry back the weighty
shellac pressings of my Signature 78 rpm singles.

In subsequent years and decades, Frank Military would figure
most crucially in my career. Among many examples, and as I have
already mentioned, it was his heroic efforts that helped save the
original Louis Armstrong "What a Wonderful World" recording
session.

I am also grateful to Frank Military for my first popular hit
artist, Alan Dale. Frank insisted I listen to Alan, at the time, in
1947, when he was a vocalist under contract to Ray Bloch on the
very successful Ray Bloch & His Orchestra radio program. (Ray
Bloch was an already famed radio personality who would, in the
1950s and '60s, achieve even greater national renown as the con-
ductor for both the Jackie Gleason and Ed Sullivan landmark television variety shows.) Dale was doing a solo stint every Sunday at
Leon & Eddie's, an illustrious Manhattan restaurant that catered to
the show business crowd and where Frank would regularly hang
out with his friends Dean Martin and a young comedy comer, Alan
King.

As soon as I heard Alan Dale, I immediately wanted to sign
him to my Signature Records company, until that time primarily a
jazz label. It turned out I had to sign the Ray Bloch Orchestra to
Signature for Dale to be on the label, and within a couple of
months, Alan, Ray Bloch, and Signature Records all had their first

big pop hit: "Kate (Have I Come Too Early or Too Late)," a tune that went all the way up the charts to #11 in the fall of 1947. The next year, with Connie Haines and "Ray Bloch's Swing Eight," Alan Dale and I had another hit, "The Darktown Strutters' Ball."

After Signature Records folded, Alan had his own television variety show, and then, when I began my Coral Records association in 1953, I was Alan's producer for a string of successes: the Top 10 "Heart of My Heart" with Dale, Don Cornell, and Johnny Desmond; "East Side, West Side (The Sidewalks of New York)" that teamed Alan with Desmond and Buddy Greco; and then the enormously popular Alan Dale 1955 solo singles, "Cherry Pink (and Apple Blossom White)" and its even bigger hit sequel, "Sweet and Gentle."

I was still running Signature Records when I planned and perpetrated my felonious abduction of Frank Military. I should explain that the New York music community was a close-knit brotherhood in those days. There were so few of us in comparison with the present-day industry that record producers, music publishers, song-pluggers, disc jockeys, and artists would always hang out at nightclubs and record sessions together when we weren't eating, celebrating, or running errands in each other's company. Too, the members of our fraternity were all young, more professionally successful than we could have imagined or some of us were prepared to be, and as a result, many of us were cheerfully, continuously, and arrogantly contemptuous of most corporate norms and societal conventions.

Frank was at my office one sun-bright morning when I asked if he would like to spend the day with me while I drove around to visit a few record distributors and sell the current Signature catalog, after which we would go somewhere for dinner.

The trunk and back seat of my car were fully loaded with quantities of shiny, new Signature singles when I met Frank downstairs from my office, and we began several genially carefree hours tooling around the city gossiping, selling records, and collecting the cash for them. As we were having a good time, I then suggested we visit some distributors in New Jersey, a notion Frank innocently and most amiably accepted.

He first expressed a small concern when, after a few further local transactions, we took off in a direction away from the Lincoln Tunnel and were speeding on the New Jersey Turnpike toward points obviously beyond Metropolitan Area parameters. Although his protests increased with each ignored turnpike exit, I smilingly made jokes and other unrelated small talk until we arrived at the warehouse of the Allentown, Pennsylvania, record distributor. It was there that I informed my by now quite disconcerted chum that it was obvious he badly needed a vacation, and since I wouldn't mind some time off myself, a few weeks in, say, New Orleans was exactly what we required and deserved.

Whether Frank felt I was correct, or could not figure out how to get away from me, or, most likely, realized he was inextricably ensnared with a possibly dangerous and unquestionably deranged companion, he agreed. Using the by now large cash proceeds from my record sales of that day, and later on supplemented by funds from my father in New York whenever I telephoned to request further infusions of money for my instantly actualized "business trip"—my dad had long since stopped questioning his son's music career (and besides, my business was now doing better than his was)—off we went on our trip way down yonder. (We visited and conducted additional business with several record distributors along the way in the South. They were enormously astonished and appreciative two "bigs" from New York would actually show up in the provinces to see them.)

We next spent almost three completely sybaritic and frequently comatose weeks at the Marmaduke Hotel in New Orleans before increasingly mounting home office pressures forced Frank Military (by now most regretfully) to drive my car back to New York while I flew west, where I spent the next two weeks at a Hollywood hotel as a violently ill tenant recovering from my prior Big Easy excesses.

Frank revealed the cause for his disappearance of so many weeks from the New York scene to very few people when he returned. Although he enjoyed the most trust and credibility of any music man in the business, even he knew no one would believe the true story!

6 Business Weak

I

IN the record business, the small independent label is the music industry equivalent of the 1940s movie cliché, "Golly, gee, let's find a barn and put on a show!" And like the characters played by Judy Garland, Mickey Rooney, and Donald O'Connor in those countless films, most independent record companies—especially, it seems, jazz labels—are invariably begun by naïvely optimistic, underfinanced, inexperienced owners whose unavoidable doom is hastened, also seemingly, in direct proportion to their depth of dedication to the music and dearth of knowledge about the industry.

As was my experience with Signature Records, anyone can make a record and press and release it for a few thousand dollars. Over the years I came to learn that the amount of money you have to pay to be realistically competitive in record production, sonic standards, marketing effectiveness, and graphics *plus* the disbursement of royalty payments to both music publishers (who pay the composers and lyricists) and artists make it practically impossible for most independent labels to survive. Before I associated exclusively with the majors—RCA, CBS, MCA, etc.—all of my independent labels lasted no more than three years, despite whatever and considerable successes we generated. This was even true with Hanover-Signature, the flourishing company Steve Allen and I founded.

That label, my last independent venture, was a particular disappointment. We had great success with the now legendary recording of Jack Kerouac—backed by jazz saxophonists Al Cohn and Zoot Sims—reading his poetry. Additionally, with stars from Steve Allen's television show such as Bill Dana ("Jose Jimenez"), Pat Harrington, Jr. ("Guido Panzini"), Louie Nye, Don Adams, and Steve himself, Hanover-Signature was among the very first companies to release hit comedy albums. We were also extremely proud of our jazz catalog with such giants as Toots Thielemans and Ray Bryant (who had a big single, "Little Susie," with us), and with bebop-scatting squirrels portrayed by music greats Don Elliot and Sasha Berland, we had a *huge* hit called "The Nutty Squirrels" that sold close to a million copies.

The type of distribution you have determines whether you are an independent label (distributed by separate sales companies in separate geographical territories) or a major label (distributed by a national or international conglomerate with internally owned sales branches in each territory). What continues to be an insurmountably unfortunate fact of record business life is that whoever controls the distribution, controls the cash flow, and independent distributors, who monopolistically control that primary life-line for independent labels, are, instinctively and habitually, killer parasites.

The record business was built by independent labels and distributors since, at the dawn of the industry, there was no one else. When, inevitably, majors began to proliferate, the independent distributors were left with those labels that remained unaffiliated and that, for the most part—Motown and A & M Records were notable exceptions—were understaffed and indigent.

To survive, the independent distributors intuitively determined that they would indeed continue to profitably sell records to retailers, but would only pay labels based on the ability of those labels to produce more product for them. With devices such as 60- (if you have any clout at all!), 90- and, 120-day billing, unlimited return provisions, and the all-time classic "contingency reserves," a label would never be paid for current sales. They were paid only for past activity when they released new recordings (and in fact when the distributor felt like paying), and frequently would go belly up before "the check was in the mail." It is still not uncommon for an independent label to have a hit, and then, after the huge investment in promotion, pressings, advertising, and artist support costs required to achieve hit sales, to go out of business before its distributor (who can always sense when a label no longer has the finances to survive or continue to produce) has to pay them.

Independent distributors have never cared if labels remain alive, for they arrogantly believed that they would always survive while the vulnerable record labels they exploited came and went. The fallacy of this myopic approach can now be seen in the increased bankruptcies of independent distributors, as independent labels are increasingly viewed as poor investments. Thus they decrease in numbers, while, as a consequence, the industry is further dominated and defined by the few mega-conglomerates and the labels and distribution networks they own. As with the dinosaurs they increasingly resembled, the independent distributors self-destructed in their cynical presumption that independent labels had to perish so they could live. Thus, tragically, the small independents that provided invaluable excitement and creativity and gave birth to a vibrant industry are gasping to an inescapable extinction.

II

And then there are record artists, most of whom believe their company is "ripping them off." In actuality, and in *every* business, "ripped off" occurs only to the extent that a lack of professionalism and/or awareness of industry realities permits. Competition is more intense and sophisticated for the attention and commitment of industry powers than at any previous time, and all too many creative artists are happy to conveniently place blame elsewhere in order to camouflage their own career deficiencies.

Record companies are fully aware that knowledgeable creative talents are the basic essential to their corporate health, and the fact is that enduring career success is proportionate to the amount of acquired knowledge necessary to compete in the arena of your industry with other business professionals. Steve Allen,

Lawrence Welk, Duke Ellington, Quincy Jones, and, more recently, Wynton Marsalis, Barbra Streisand, Billy Joel, Paul Simon,
etc. are all spectacular successes, not only for their incomparable
artistic abilities, but because they made sure they knew more
about how their industry functions than their rivals. This enables
them to become controlling owners of their businesses, hire accomplished professionals to work for them, and, because they
were at least as aware of all aspects of their industry as their
"expert" employees, they are able to demand and obtain the highest level of achievement so as to maximize the potential of their
own creative talents.

When I am invited to speak at college seminars, I plead with
musicians to understand that, while half the battle was always to
become a professional musician, the other and vastly more important goal today is to be a *music professional*. For their sakes and
for the music, I hope they comprehend this crucial advice. Music
would be better, and a much larger public would support it.

7 Coral Riffs

I

CORAL Records began as an attack missile, but, I can say with some pride, it was not a bomb.

After my Signature label folded, both the master recordings and myself knocked around the business for a while before ending up at Decca Records, where I was able to transfer the Signature catalog, and also be hired as an assistant producer at the newly formed Coral label.

Jack Kapp, the prideful and imperious president of Decca, was furious at the independent labels and distributors that were rapidly re-emerging at the time. Kapp felt the independents—who, in fact, had founded the record business—were parasitical squatters who would deplete the exclusive marketplace that had been expanded by the major record labels and was now their sacrosanct dominion.

The plan was to insinuate into the commercial arena an all-purpose, independently distributed label—with major label budgets and staffing—whose competitive superiority would consequently run the hated homesteaders off the land. Decades later, when the most immense expansion in record business history was at its

peak, the major labels once again attempted to eliminate the independents. This time, in their hubristic haste to accomplish the
deed, the corporate cabal overlooked the fact that the independents were the labels that developed the new mega-star artists and
producers the majors invariably lured away and on whom, in truth,
they were dependent. The fracas quickly resulted in two prevailing
staples of the present-day record industry: the continuously accel-
erating rise of successful new independent labels and the inexorable drive of each major to absorb all of these profitable upstarts
before their mega-conglomerate competitors did the same.

Back, however, in the industry's more modest days, the artistic area at Coral Records was run by Milt Gabler, an old friend and
respected rival. When I was a teenaged record business novice
peddling my Signature pressings to every New York City record
store that would take them, Milt owned one of the very first successful jazz record stores—the Commodore Record Shop—and,
thankfully, I was among his best customers. Then Gabler turned
a legendary jazz store into an even more legendary jazz label,
Commodore Records, which eventually traveled the same road to
extinction as Signature Records, and similarly followed its father
to the corporate embrace of the Decca company.

Milt Gabler was transferred to Decca soon after I arrived at
the corporation, and I was appointed to run the smaller Coral
division and chase the independent companies out of the business.
I suppose I was the logical "anti-establishment" sort for the damage
my job mandate required. Decca, like its companion competitors
RCA and Columbia, was one of the venerable firms that comprised
the "ruling class" of the record industry. It was the hallowed home
to such American Music fixtures as the Andrew Sisters, Bing
Crosby (his Decca "White Christmas" is still the largest-selling single record of all time), the Mills Brothers, Guy Lombardo, and
Lawrence Welk. Also, as its record industry counterparts and the
show business power hierarchy in general, Decca was comman-
deered by rough-hewn gentlemen from an era previous to mine.

One instance of how unpolished were my Decca superiors
involved a party given by the president of Decca Records, Milton
Rackmil, for the executives of the prestigious German classical
music label, Deutsche Grammophon. The German company, after
many months of difficult negotiation, had agreed to extend their
American distribution deal with Decca for another three years.

The Second World War had been over for almost a decade, and among the commercial exchanges between former enemies was an arrangement whereby American Decca would manufacture and distribute Deutsche Grammophon in the United States. During the initial term of the contract, the Germans became extremely displeased with the quality of the American pressings, and indicated they would only renew if it was ageed they could manufacture in Germany for export to the United States market.

With the dispute resolved, and a mutually advantageous relationship salvaged, American Decca hosted a gala occasion in the penthouse suites at the Sherry-Netherland Hotel on New York City's Fifth Avenue exclusively for the most senior executives of both companies. Although the Coral Records division was a small subsidiary of the Decca Corporation, as its current chief, I was told to attend.

Arriving at the designated suite, I was abruptly thrust into a scene even more grotesque than any I could have imagined. The Germans were red-faced, staring straight ahead, stiffly seated together in mortified silence, while all around them the Decca executives were uniformly intoxicated. I swear this is true: they were goosestepping, regaling each other with "Heil Hitler" salutes and defiant arm movements, pigeon-German, and every other bit of adolescent grossness they could perform for each other. It was Judgment at Nuremburg in the Twilight Zone.

I couldn't believe it, and, after a frantic search, I found a drunken Milton Rackmil sequestered in a private, adjoining suite. When I expressed my dismay and pleaded, "Rack, please, you have to do something," he refused to move (perhaps he couldn't) and replied, "Who gives a shit. We got those Krauts for three more years, and they can go screw themselves. That Curt Kienkle (a Deutsche Grammophon senior vice president) in there was a U-boat commander. Fuck 'em all!" As I went back to this party from hell, the contemptuous Germans rose as one body and, without a word or sideways glance, swiftly exited this demented affair while no one at Decca took notice of their departure.

II

When my contemporaries and I arrived at Decca, we found a brash and brassy Garment Center crowd with their striped suits, pinky rings, cigars, and watch chains while we soft-spoken suburban sophisticates smoked pipes, dressed in sport jackets and loafers, and wore sunglasses indoors. It was as if Thelonious Monk was playing piano duets with Eddie Duchin, and it was the right time and the right place for me.

From a pop perspective, these would be my glory days. For nearly a decade, I had hit after hit flying out of there. It was unreal. I was suddenly the young genius at Decca Records, a prolific "hit-meister" who could do no wrong.

It was also an advantage to work with such a panoply of experienced performers. Debbie Reynolds, the movie star, had her only #1 record, "Tammy" (which topped the *Billboard* magazine lists for five weeks straight) on the charts for over half a year. My friend Lawrence Welk hit #5 when we made "Oh Happy Day," and around the same time the great Pearl Bailey charted in the Top 10 with "Takes Two To Tango," her only hit single.

My association with the McGuire Sisters produced over thirty hits that included "Goodnight, Sweetheart, Goodnight," "Muskrat Ramble," "Something's Gotta Give," "He," "The Theme from *Picnic*," and two #1 singles, "Sincerely" and "Sugartime," both on the charts for about half a year each.

The popularity of the McGuires was equaled only by that of my future-to-be-wife, Teresa Brewer. We were both married to others when we worked together at Coral, but it seemed that every time we were in the studio together, we would make hit records.

With such prior hit single successes as "Music, Music, Music" and "Choo'n Gum," Teresa was already a major recording star when we both arrived at Coral and began our professional relationship. Almost immediately we hit #1 with "Till I Waltz Again with You," which held that position for the seven weeks of the half-year it remained on the singles charts. "Waltz" was followed with a consistent string of hits (a total of 35) that included "Ricochet Romance," "Jilted," "Let Me Go, Lover!," "A Tear Fell," "A Sweet Old Fashioned Girl," and "You Send Me"—all Top 10 singles.

Teresa was also so versatile that I was able to team her with everyone from Don Cornell to Les Brown to Mickey Mantle (!) and have hits each time. (Much later, after we were married, I remembered that Teresa's first huge hit, "Music, Music, Music" was recorded with my old Dixieland jazz chums, Max Kaminsky and Cutty Marshall, so I began to record her with *all* my jazz friends.

The albums with Duke Ellington, Benny Carter, Dizzy Gillespie, Count Basie, Stephane Grappelli, Earl "Fatha" Hines, and, most recently, Wynton Marsalis and David Murray, have been acclaimed by critics the world over, found a new audience for her, and established Teresa as a major jazz artist.)

At Coral Records, and thanks to Teresa and all the other artists, I was *the* young "hot" producer. The brass wanted me to be happy, so I was even allowed to record jazz with such greats as Terry Gibbs, Manny Albam, and Hot Lips Page, as long as the hits did not stop.

One of those hits involved a sweet revenge. When I first joined Coral, I brought over Alan Dale, one of the most popular singers from my recently bankrupted Signature Records company, to the label. In order to quickly establish Alan at our new home, I had the idea to combine him with two other successful Coral vocalists, Johnny Desmond and Don Cornell, in a recording of the old tune "Heart of My Heart." (I was always thinking up "odd" combinations of artists and songs to record that would, somehow, work out sensationally.)

Two of Alan's biggest hits for Signature were "The Darktown Strutters' Ball," in a duet with Connie Haines, and the Italian folk lament, "Oh Marie." He would soon have another big hit—this time on Coral—when I produced his record of "East Side, West Side" with Desmond and Buddy Greco. Years later came such other unorthodox pairings as "What a Wonderful World" for Louis Armstrong, the John Coltrane *Ballads* and Johnny Hartman albums, and then, as previously mentioned, Teresa Brewer with everyone from Mickey Mantle to Count Basie.

When my plans for "Heart of My Heart" were made known inside the company, a champion of my budding record business career robbed me. Milt Gabler, now the head of artists and repertoire for Decca Records, heard about what his protégé was about to do with the three male vocalists on Coral, and, in sudden appreciation of how perfect the song would be for *his* big Decca act, the Four Aces, he went into the studio with them and literally overnight "aced" out my record. This was one time Goliath demolished David.

Retribution occurred on the West Coast some time later. For around two months of every year I was employed at Decca as head of artists and repertoire with Coral Records, and I would

record and work out of the "state-of-the-art" Decca studios on
Melrose Avenue in Los Angeles. Their equipment and staff were
superb, and it was always refreshing to experience the relaxed
but nonetheless unexcelled professionalism of my California colleagues.

One afternoon, with a few hours of "down time" available to
me, I went to see the film *Picnic,* and when I heard the treatment
of the song standard "Moonglow" with that extraordinary violin
counter-melody, I "freaked out" (as the rockers would say).

I got to the office early the next morning, and immediately
ordered the soundtrack—I had to hear this again. *Picnic* was produced by Universal Pictures, another division of the corporate
behemoth that also owned Coral and Decca Records. I began to
formulate what I believed would be (this time) an easy, uncompli-
cated in-house collaboration, despite past experiences of intense
competition between the authoritarian Decca Records division and
the younger, more contemporary Coral Records subsidiary they
owned.

Joseph Gershenson, the Universal music chief who, as a matter of course, recorded soundtrack albums for Decca, considered
the "Moonglow" juxtaposition as a "throwaway" gimmick in the
score composed by George Dunning, and doubted whether anyone
would be enthusiastic about this strictly background music. (Movie
people called this category "underscoring.") Then, when he (and
quickly, Decca) learned that their "hot" Coral Records pop producer wanted to release a single called "Moonglow and the Theme
from *Picnic,*" Gershenson's competitiveness was ignited, and Coral
was informed to desist, as Gershenson and Decca were rushing out
the single from the soundtrack album.

I never moved so fast in my career. Before the day had ended,
I had an arrangement written and copied, and recorded my
"Moonglow/Picnic" single with George Cates and His Orchestra
(which was mostly the Lawrence Welk television show orchestra,
for which Cates was the musical director, and who, with Welk,
was one of the Coral artists I produced). Both recordings were
simultaneously released: the soundtrack single on Decca, and my
"renegade" on Coral. Despite the tremendous "in-house" rivalries
that predictably ensued, both singles did very well. Decca (naturally) sold a million copies, and the nesters at Coral did over a
half-million.

Then the music publisher, aware that the recordings had matched the film's success—and also of the rivalry at Decca—telephoned me to say they were about to put lyrics to the *Picnic* theme, and would I be interested in recording it again. I continued to love this music, and I told Al Gallico at Shapiro-Bernstein Music, "Absolutely. Let me know as soon as the lyrics are completed."

Coincidentally, Steve Allen, a friend and Coral artist I was also producing, was as intrigued with the theme as I was. As always an enterprising renaissance man bursting with literary ambitions he would soon successfully realize, Steve told me, "Bob, I'd love to write these lyrics."

I couldn't wait to telephone Al Gallico with this news. To avoid any possible impediment to this most felicitous marriage of music and lyrics, I informed Al, "I don't know who you have to write the *Picnic* theme lyrics, but if you let Steve Allen do them, I'll record it with the McGuire Sisters." Only the hottest act at Coral Records that year. Gallico unhesitatingly said, "You got it!"

We never found out which lyricist at Shapiro-Bernstein lost a chance at a huge hit single when the McGuires made a greatly successful record called "The Theme from *Picnic*" with music by the well-established film composer George Dunning and words by a (then) neophyte lyricist named Steve Allen. Moreover, I had my second big hit in three months with a song my corporate superiors would rather I had not produced even once!

III

The Decca studios on Melrose Avenue in Los Angeles was the locale for another, and at the time, highly unusual hit record that I again had to convince the corporate lords at Decca/Coral to release.

Dave's Blue Room on La Cieniga Boulevard was a restaurant/nightclub around the corner from the studios where everyone would eat, have meetings, and socialize. I was having dinner there one evening when, to my delight, I learned that the comedian Buddy Hackett was about to perform an unscheduled show in preparation for some upcoming television appearances in the city.

One of the newer routines I heard was a riotous sketch about a Chinese waiter with scarcely disguised hostility taking an order from a table of tourists. This was years before Bill Cosby, Bob Newhart, and the stars of the Steve Allen Show—Bill (Jose Jimenez) Dana, Louis Nye, Don Knotts, Tom Poston—had major hit comedy albums, but I was convinced this material, with some background music, would be a hugely popular single.

With a career that always thrived in nightclubs, resorts, and, more recently, films, Buddy Hackett did not believe that he could make a hit record at the time I introduced myself and proposed we go into the studio to record his Oriental waiter routine. Still skeptical, he agreed to accept the token $300 fee against a 5 percent artist royalty I offered and show up the following afternoon.

The next day, Buddy had not arrived after the first hour scheduled for the session. When I telephoned his room at the Sunset Continental Hotel, a drowsy-voiced Hackett responded with, "I thought you were full of crap and this was just a practical joke. Anyway, who would seriously want to record this shit?"

When I informed Buddy that three engineers and an anxious record executive were waiting for him with the utmost seriousness, he soon appeared, and we made a record that eventually became one of the very first comedy hits of the decade.

Back at New York headquarters, predictably, my bosses were extremely apprehensive. While they thought it was convulsively funny and would unhesitatingly play it for friends in the privacy of their offices, I was told Coral was a label with musical artists all America loved and wanted their children to emulate. Of course Buddy's "Chinese Waiter" was fine for sophisticated New York corporate executives, but the great unwashed, all-American Coral Record audience would *never* accept such an assault on their sensibilities.

Fortunately my prior passions as an obsessive teenage recording scholar provided the remedy. I reminded my chiefs that as far back as the 1930s, ethnic satires with music—such as Henry Burbig's Germanic burlesque to the *William Tell Overture* on RCA, along with all the Top 20s Spike Jones had with the same label ("Der Fuehrer's Face," "The Sheik of Araby," "Hawaiian War Chant," and even in 1950, "Chinese Mule Train")—had been immensely well received by the very same middle-American proletariat that Coral Records pursued.

Soon after, I had among the first of many hit comedy singles at the label, and Buddy Hackett, from his trifling advance and royalty, pocketed well over $30,000, in those days a staggering sum for an artist to earn from record sales.

The next time I saw Buddy Hackett was in Manhattan during the 1954 Christmas season when he visited a recording session I was producing at Coral with Alan Dale that became Dale's first Top 20 single for the label. The song was "Cherry Pink (and Apple Blossom White)," and little did we know when we started that day how prophetic that title would become an hour later. Thanks to Buddy, who was Alan's old friend and fellow native from the same Brooklyn neighborhood, we barely got through the session.

That winter was among the coldest in memory or on record, and that afternoon Hackett entered the studio wearing the largest, most suffocatingly thick fur coat anyone had ever seen. After he was satisfied everyone had appreciated his grand entrance and outrageous garment, Buddy sat down in front of Alan and the orchestra looking like an obese, furry basketball.

Naturally, Alan invited his guest to take off his coat, and Buddy, ever irrepressibly irreverent, promptly stood up to reveal that he was completely nude!

Pandemonium! Engineers, musicians, secretaries, executives, and one helpless record producer were now uncontrollably hysterical when they saw this vision of a major show-business celebrity in all of his cherry-pinked glory. Nor did it end there. Since Buddy claimed he would never leave his old pal at "such an important career moment," and in fact had nothing else to wear and "it's very warm in here," he sat outright naked in the studio for the rest of the session, uninterruptedly mugging. With our sides aching and weak from laughter, we somehow (dare I again say "barely"), managed to finish a hit single that was among the most entertaining and simultaneously terrifying experiences I ever had in a recording studio.

8 Buddy Building

I

IT was also at Coral Records that I made two most famous discoveries. In the music business, though, it is sometimes difficult to determine who discovered whom.

At Coral, I had a recording budget, so many dollars per fiscal quarter. Every record that was made, every song, every artist all came under the jurisdiction of just one executive in the company—the head of artists and repertoire—me. The president of the company didn't care who you recorded or what songs you did, as long as you were selling enough records to make a profit every three months. (It is completely different today when the first person to consider a new record may well be an attorney, a division administrator, or a bean-counting controller. Then the record is played at a big meeting that involves maybe nine department heads: sales, marketing, lawyers, accountants, everybody but music people. I am often amazed records actually get out of these corporations and to the public.)

As I remember, Murray Deutch, an old friend who was the general professional manager of the powerful Southern Music publishing house at the time, visited me with some recordings of a group of unknowns from New Mexico called Buddy Holly and the Crickets. In those days, music publishers were the primary agents for record deals. In exchange for the publishing rights, they would secure record releases and contracts for the groups and songwriters they felt had potential. Two years earlier, for example, Murray had brought me a song for Alan Dale to follow our big hit "Cherry Pink (and Apple Blossom White)" on Coral. Alan's recording of "Sweet and Gentle" became an enormous smash and the last Top 10 single I produced with him.

Murray had received the dubs from a Clovis (New Mexico) record producer named Norman Petty, thought they were terrific, and took them around to the big major labels—Columbia, Decca, RCA—all of whom thought this early meld of rock 'n' roll and country music sounded too impossibly dumb and crude for any public acceptance.

After all the geniuses at the bigs had rejected it, it was time to give the second-class citizenry a shot. Jerry Wexler, chief of the gutsy and at the time independent Atlantic Recording Corporation, listened to Holly and the Crickets and reacted as negatively as his colleagues at the major companies. Coral Records was next on Murray Deutch's B list as it was basically a small-label subsidiary of the massive Decca Corporation (who had previously and very firmly passed on the group).

When Murray played the recordings for me, I became rabid as soon as I heard "That'll Be the Day." I said, "Jesus Christ, this is fantastic," and wanted to release it immediately. It would only cost a trifling $2500 (small change even then—not even tip money now) to buy the master, and, as excited as I was, I anticipated some problems. I knew my production budget for the present fiscal period was depleted, and that my bosses—as rigidly unimaginative as they were miserly—could not possibly comprehend what Buddy Holly and his group had originated.

Since I could not immediately give Murray a genuine commitment, he continued to solicit the other smaller labels. As the unceasing rejections multiplied, Murray would frequently return to me, the only record executive who had shown any interest at all, both to learn if my proselytizing within Coral had changed attitudes and to reaffirm he wasn't as crazy as everyone else kept telling him he must be to persist with this "hillbilly garbage."

Meanwhile, I was enthusiastically playing Holly and the Crickets for a lot of people in the company, and everybody again said, "Forget it! You can't put this record out." The president of the company told me, "This record absolutely cannot come out on Coral. It'll destroy the image of the label. We have great and beloved artists like Lawrence Welk, Debbie Reynolds, the McGuire Sisters, Teresa Brewer, and now you want to come out with this horrible music by something called the Crickets."

Just the same, as with the Buddy Hackett situation previously, I was fortunate to be a discophile. I finally remembered the corporation owned an obscure record label named Brunswick, and that solved the problem. Brunswick was devoted to "race music"—a prior euphemism for rhythm and blues—so I went back in and said, "We have to release this record, and if all of you are so concerned about the image of Coral, put it out on the Brunswick label. $2500, what can we lose?"

Offered this simple, face-saving solution, the company agreed to buy and release "That'll Be the Day." I also believe that they let me put it out for another quite simple reason: They wanted to keep me happy. I was of value to them, giving them a seemingly uninterrupted parade of hit records, and they didn't want to really upset me. Besides, and unlike most of the stars I was producing, I was full of my own success and power, very emotional, arrogantly rebellious, and ready to quit whenever my authority was questioned. I'd always regret it and apologize, but I'd go through the motions.

One time, as an example, after I had produced the theme music recording for "The Milkman's Matinee" radio program gratis as an acknowledgment courtesy for New York City station WNEW, then the most influential pop music broadcaster in the country, and its "Milkman," Art Ford, I destroyed the only station copy of the theme in an impulsively rash moment when I felt Ford showed lack of respect for me. As all of us in the industry were so entwined with each other in those days, it was customary—and encouraged—for music executives to make unannounced visits to WNEW at any time to trade gossip and pal around with the station on-air personalities and management. On the occasion of this particular visit to "The Milkman's Matinee" show in the middle of the night with my friend Frank Military, a distracted Art Ford did not, as was usual, effusively greet us as visiting monarchs when we entered his studio. As a Mr. Big Record Producer, in my arrogance and

perhaps under the influence of the spirits I had drunk before coming to the studio, I decided to be instantly insulted and then con-
temptuously broke the shellac recording of the theme and angrily
took my leave. (I cannot guess what was used to sign off the show
that morning, but I would safely bet it wasn't one of my records.)
Frank Military still doesn't believe I committed the outrage he
witnessed, and the light of day brought my embarrassed remorse
and, at my expense, a new copy of "The Milkman's Matinee"
theme delivered to the station library in time for the next show.

I was never as confident in my life anything would be a hit
record as I was about "That'll Be the Day." A few weeks after it
was released, we were at a convention in the Midwest somewhere.
We used to check in with the New York office every day, and this
one time the sales manager said, "Bob, I don't know what happened, but the distributor in Philadelphia has re-ordered 16,000
copies of the Buddy Holly to be shipped to him overnight." Then
the record zoomed to #1, and everybody went crazy.

No longer anonymous hicks from some Southwestern state
nobody ever went to, Buddy Holly and the Crickets had become,
in industry parlance, "monsters." "That'll Be the Day" would be
their biggest hit, staying on the charts for half a year.

Now that it was suddenly and very profitably permissible for
Buddy Holly to join the roster of mainstream pop stars on the
Coral label, my grateful bosses accepted another idea I had. It was
originally my "plan B" if they would not permit the Holly record
release on Brunswick. To cash in on two different entities, from
that point onward, Buddy Holly would be on Coral, and the Crickets would be Brunswick artists even though they were one group.
The wisdom of this arrangement was immediately apparent when
their next two (and second and third largest) singles, "Peggy Sue"
by Buddy Holly on Coral and "Oh, Boy" by the Crickets on Brunswick, were simultaneously released a couple of months after
"That'll Be the Day." Both hit #1 and had coexistent six-month
chart runs.

I was really thinking commercially. Buddy was the personality, and we really didn't want to bust up the Crickets and Buddy
Holly. I felt we'd get more exposure, more of a run for our money,
by having the music out two ways. The decision about who would
record what was left to their main producer, manager and guru
Norman Petty, the man who had found Buddy and the group in

the Clovis, New Mexico, area where they all lived. Back in New York, we'd say, "We want a Buddy Holly record for the next release, and then we'd like a Crickets record." Norman Petty would go in the studio he had in Clovis, and make the records virtually on order.

Buddy Holly, the group, and I didn't actually meet until after the third hit record. Murray Deutch and myself were invited to Clovis, which was about 90 miles from Buddy's hometown, Lubbock, Texas, to be honored, receive keys to the city, cowboy hats, whatever. It was weird. When we walked off the plane, you would think it was Lyndon Johnson coming home. We stayed for three days and got to know everybody.

Buddy and I talked quite a bit, and got to know one another pretty well. He was a terrific young man, very unassuming, and a real gentleman. He always had a suit and tie on, so opposite to what subsequently became the record star norm. (Some time later, Buddy would marry Murray Deutch's secretary.)

My new friend was so grateful that I had fought as hard as I did for "That'll Be the Day" to be released, that he said, "Bob, I gotta really do something for you. Write a song, willya, for our next single?" So, I called these two friends of mine, Ruth Roberts and Bill Katz, with whom I had co-written the lyrics to Duke Ellington's "C-Jam Blues" (which was then retitled "Duke's Place," and was since recorded by many singers including—separately and together—Ella Fitzgerald and Louis Armstrong, most famously).

For the new Holly single, we wrote "Mailman, Bring Me No More Blues," probably the most simplistic blues anyone ever heard. It ended up as the B side of one of the few unsuccessful Buddy Holly singles—the A side was "Words of Love"—and both sides were recorded years later by a then new group, the Beatles, who always mentioned the records of Buddy Holly and the Crickets as among their favorites and main influences.

It was interesting that I had to use an alias—Stanley Clayton— as a co-writer of the song Buddy Holly specifically and openly had asked me for. There is nothing to hide anymore, but in those days there were all sorts of shady deals everywhere, and the producer of a record was not supposed to record or publish his own music in his own name. "Conflict of interest," which was really a crock. The brass of the company knew. Everybody always knew. I came down in the elevator one day, and a Decca Records vice president says to me, "Boy, that Stanley Clayton sure has a lot of songs with us." He knew, and he was letting me know he knew. It was never really a lot of money anyway, and as A & R men were not among the more highly compensated staffers in the organization, we were encouraged to participate, pseudonymously, in these evasions.

Despite the fact that Norman Petty had an iron grip on every creative and business aspect of Buddy Holly and the Crickets, he was jealous that Buddy also wanted to record in New York with me as producer. We all had one of the most successful associations the music industry had ever seen, and year after year Buddy would tell me how much he wanted to work with me in a New York studio. Some last-minute "something" would always postpone this ambition (as previously mentioned, there were shady deals everywhere), but finally, in the last year of Buddy's life, we got together to make just two sides.

Several accounts have erroneously stated that the date wouldn't have happened unless I agreed that Norman Petty could play piano on the session. I did agree we would record one of his tunes, but Petty was never present at the actual date.

Time and musicians were booked at one of the hottest and most distinctive studios of the era. Bell Sound Studios was located on Eighth Avenue in Manhattan. Unlike the traditionally configured studios at Decca, or anywhere else including the Clovis, New Mexico, command post, it featured a revolutionary "dead sound"—sound that wasn't traveling all over the room—which allowed producers to isolate the individual musicians and achieve a literally unheard-of depth of sonic separation, vibrancy, and excitement. I knew it was where I should record Buddy Holly.

I will always be proud and grateful that the only time I was hands-on, emotionally involved with Buddy Holly in the studio from start to end, we produced a magnificent record. We did the standard "That's My Desire" and Norman Petty's tune "Rave On." The chemistry was perfect, and Buddy was so invigorated to be at last recording in *the* New York City studio, with the cream of the city's session musicians, that he gave one of the performances of his lifetime. "Rave On" stayed on the charts for ten weeks, and is still considered both a Buddy Holly and rock 'n' roll classic.

II

There was certainly one other aftermath of the visit Murray
Deutch and I made to Clovis that should be mentioned. I'm sitting
on the veranda fence at, I think it was, Norman Petty's house, and
the Crickets' lead guitarist, Sonny Curtis, like all musicians who get
to meet a record producer, tells me he's dying to have some of his
songs recorded. I wanted to be polite, especially to a member of
the biggest group I had on the label, and there was obviously not
much to do in this town, so I agreed to listen.

After a couple of songs, he plays and sings "Sugar in the
morning, sugar in the evening, sugar at suppertime." I said, "Don't
play that for anybody else. When I get back to New York, I'm
recording it with the McGuire Sisters." So he says, "Okay." Much
later I found out that my pal Murray Deutch, who was always one
of the best in the music business at recognizing new talent, had
already signed Sonny Curtis to a contract as a songwriter. (It would
still be a few years until Sonny Curtis wrote his own lifetime annu-
ity: the *Mary Tyler Moore Show* theme song.) Murray's astute
move was never a problem for me, and fair is fair. If Murray
Deutch hadn't asked me to hear those cornpatch dubs by an unknown group and producer from a town in the boonies I never
heard of, I never would have "discovered" Buddy Holly.

"Sugartime," in fact, provided a remarkably interesting view
of the pop recording process in those days. I had become extremely frustrated with how the rhythm sound was recorded. It
was never exactly the way it sounded live in the studio, which was
always more exciting than the final product. It seemed nobody
knew how to control a bass or an electric guitar or drums and the
cymbals, and every time they continually got lost when the record
was made.

Bell Sound Studios had just opened with their innovative isola-
tion capacity, and at last I was able to record the rhythm precisely
as I heard it. "Sugartime" was the ideal song for the new and
welcome recording techniques Bell Sound now made possible, and
I eagerly booked the time for the McGuires, arranger Neal Hefti,
and a sixteen-piece big band.

A few words about Neal Hefti, a solid music professional who
composed and arranged some of the most important records
Woody Herman and Count Basie made. He had played trumpet

with the Herman, Basie, and Harry James bands, composed hit songs such as "Girl Talk," as well as *The Odd Couple* and *Batman* television themes. Neal was so well respected, the most in-demand session musicians in the industry would turn down everything else to play on a Hefti record date.

Now the hottest group, producer, and musicians in the business are doing "Sugartime," and I can't believe what I'm hearing from the control room speakers. I was completely startled, confused, bothered, and bewildered by the unaccustomed clarity of every separate musical element in my ears. I kept ordering take after take, saying "not right, not right," and becoming the most despised person on the planet.

The assiduously respectful McGuire Sisters are looking at me with machete eyes like I'm this deranged stranger who in the next minute will permanently destroy their careers (and this after over twenty hit singles together). The studio musicians, renowned for their implacable professional quiescence, have started to talk back to me. And with each take, Neal Hefti's face becomes an even more bloodless vision of wounded, raging disbelief. An elliptical portrait of Dorian Gray.

Then, and very slowly, I start to get it. I'm hearing the rhythm, the heartbeat of every pop record like never before. I say, "Neal, we don't need the brass section. Why don't you send these guys home." So, he dismisses the brass players, and I am sensing he now thinks the most excruciating death would be too good for me. After a couple more takes, I said to Hefti, "We don't even need those saxophone figures you wrote. Send the saxes home." I was positive Neal Hefti would never speak to me again.

So, we finally recorded "Sugartime" with just the rhythm section. One take, and it was perfect! The McGuire Sisters had their second #1 hit, which stayed in that top slot for one entire month of the half-year it remained on the *Billboard* "Hot 100," and Neal Hefti continued to be among my dearest friends.

Years later, I was having dinner alone in Los Angeles, and a conspicuously agitated, ashen-faced Neal Hefti stalks into the restaurant. I was as delighted to see my friend as I was alarmed by his emotional condition, and invited him to join me. Neal immediately made it explicit that he would rather drink than eat or talk, so the next 45 minutes passed in labored silence. When he finally started to speak, he told me, "Bob, I was just fired from the best-paid movie job of my career." He then explained he had been musical director for the biggest film of the year, *Funny Girl,* starring, in her first movie role, Barbra Streisand, who was repeating her extremely successful Broadway triumph. Neal then related that these last weeks had made him a wreck. From the very first day, Streisand had been a no-class, screaming, ruthless, egomaniacal terror. Relentlessly insistent on getting her way with incessant "Doesn't anybody else around here know how to do anything?" tantrums, everyone a target for her screeching abuse. Finally, after she again embarrassed and insulted him in front of a hundred musicians and the director on a sound stage, all Neal could think of was to walk out, only to be fired from the picture a few minutes after. When I said, "Come on, you must have had worse experiences than this," he replied, "It was almost this bad only once before, but I didn't walk out of 'Sugartime'!"

9

More Fables
and Foibles

I

THE good times continued to roll. "Sugartime" was my third #1 single in six months. (The other two were "Tammy" by Debbie Reynolds and "That'll Be the Day.") Then it was discovery time again.

In those days, what would usually happen is that you'd get a call from a music publisher, a talent agent, or a personal manager who'd say, "I'd like to visit your office and play some music, and introduce some new talent to you." And that was our job: We saw the people in the industry who (sometimes) had credibility and could present new artists to us.

One of these artist representatives was Al Green, who had been around for years managing black performers mainly from Detroit and the Midwest. He wanted to sell me a vocal group called the Dominoes, who were at the Apollo Theater that week.

So I go uptown one night to hear the Dominoes and their young lead singer, an amateur boxer named Jackie Wilson, who, incidentally, had just replaced Clyde McPhatter for this New York appearance. I'm thinking to myself, "To hell with the Dominoes, let's take Jackie Wilson." (I admit there were a lot more cut-throat approaches to the record business in those days.) I said to Al Green, "I don't need another vocal group, but I'd like to sign the kid singing lead to the Brunswick label." The hustling managers and agents would crucify anybody, and all Al Green sees are dollar signs. In the next minute he tells me, "Yeah, who cares about the vocal group. You want Jackie, you got him!"

Green was staying at the Taft Hotel in New York, and I sent the contract over to him. After waiting a couple of days, I started to think, "Now that he has a deal, maybe he's shopping the other labels for a better contract." (Still a common practice.) So I telephone his room, and a young man answers and says, "Mr. Green died last night."

Before I can finish saying, "Oh, my God, this is terrible!," the voice on the telephone says, "But I have a contract here that I was supposed to give you this morning." That was my first meeting with a very personable, gentlemanly 19-year-old auto mechanic from Detroit named Nat Tarnopol, who became Jackie Wilson's personal manager soon after. Some time later in the fifteen-year association Jackie Wilson had at Brunswick, where he had over 50 hit singles, and long after I had left the Decca Corporation, the Brunswick label was actually given to Nat Tarnopol so he and Jackie wouldn't leave for another record company.

Among the first songs we recorded was "Lonely Teardrops." Co-written by a then completely unknown Detroit resident named Berry Gordy, it became Jackie Wilson's first Top 10 hit, and stayed on the *Billboard* charts for 21 weeks. I have no idea how we wound up with the song, and, like everyone else in New York, I had never heard of the songwriter.

As the world would soon know, Jackie Wilson was a superb rhythm and blues artist. What was different about "Lonely Teardrops" was that, except rhythmically, we didn't record him in the conventional R & B style. The basic sound, as typified by Fats Domino or Little Richard, was a rhythm section with a couple of saxophones, or one trumpet, one tenor sax, one trombone. That was always it.

What we did was use a regular sixteen-piece orchestra instead of the small combo sound, and we recorded in a big hall, Pythian Temple, which was very popular in those days for original cast album and huge lush orchestral dates. That overall sound—the

Rhythm & Blues feeling combined with a big band concept—made Jackie Wilson a gigantic "crossover" artist years before that term entered the music business lexicon. This unusual approach allowed Jackie to become one of the first recording stars who not only appealed to black audiences, which was a very specific market in those days, but could be accepted by the white market a lot more easily and quickly than if we had just made the more traditional type R & B recordings.

Much the same as Buddy Holly, Jackie Wilson was a very friendly, soft-spoken gentleman, always excited to be making records, and a pleasure to work with. We were making good records together, but I would say, in all other aspects of his career, I think Jackie was misguided. I don't believe he was treated properly, and, knowing the business as I do, I'm sure he never saw all the dollars he made, even though he got some new clothes and a new car and those kind of enticements.

Before long, the stories about Jackie Wilson's drug problems would become widespread. I always thought it was happenstance that Jackie found Berry Gordy instead of the other way around, which could have occurred a few years later when Gordy had made his Motown Records company a huge success. If Jackie had been a Motown artist, his life would certainly have been better in a guidance sense, as it was Berry Gordy, and visionaries like him, who soon ushered in what became a new approach at the time: totally controlling the personal lives of their record stars to insure that these performers would be healthy and financially secure after their music careers had ended.

It was enormous fun to find and work with such nice people and excellent artists such as Buddy Holly and Jackie Wilson, and I felt privileged to have made an important contribution to their careers. Making records with all kinds of performers is a constant joy for me. I've always said, only somewhat facetiously, if I was called tomorrow to record the greatest polka band in the United States, I'd say yes. Jazz is certainly my main passion, but to have hit records with everyone from the McGuire Sisters, Jackie Wilson, Teresa Brewer, Pearl Bailey, Buddy Holly, and Debbie Reynolds was great, and very exciting.

Additionally, being in a position to discover major talents (and get them recorded) was a tremendous professional perk. Along with Buddy Holly and Jackie Wilson at Coral, and Erroll Garner,

Alan Dale, and Gato Barbieri among so many others through the years, it would be an electrifying thrill to hear someone new and instantly know they could make a difference in the business, and I might also possibly make a difference for them.

II

Of course, it wasn't always champagne celebrations, and I have been at least as dumb as anyone else with my own "woulda-shouldacoulda" list to keep me awake at night. Marvin Hamlisch never spoke to me again after he displayed his piano style and some recently composed songs, and I informed him—this was before *The Way We Were, The Sting* film score, and *A Chorus Line*—he shouldn't seriously think he could have a career in the music business. I also told Jann Wenner that his vision of a mass-market monthly magazine based on the rock 'n' roll culture called *Rolling Stone* was the silliest idea I had ever heard of. (Jann nonetheless became a good friend, perhaps so he could—and does at every opportunity—remind me of what a genius I can be!)

When I was still at Coral, I also became aware of a marvelous musician, formally an arranger with Glenn Miller, who was the assistant music director at Universal Pictures under the prominent industry eminence Joseph Gershenson. His name was Henry Mancini, and he had just finished supervising the music for, appropriately, *The Glenn Miller Story*, which had started quite a "buzz" throughout the music community.

I thought Mancini was among the most creative, interesting musicians I had heard. Sure as I was that he had huge potential to achieve a broad-based popularity with record consumers, I convinced the check-signers at Coral to give him a minimum contract. We then recorded four instrumental sides with a large orchestra that laid an even larger egg.

As the deal did not involve an expensive commitment to Mancini from the company, and my enthusiasm for Henry's talent had grown as we worked together and became friendly, I wanted to try again. My bosses at Coral, transfixed with many successful hits by vocalists (including the many I had produced), didn't want to hear about a contract extension: "Forget it, he's a loser. Let's just stick to what we know will sell," and dropped him faster than Sonny Liston was dropped by Cassius Clay in Lewiston.

Mancini soon after signed with RCA, and his next record was *Music From Peter Gunn,* which was the #1 album in the country for ten weeks.

Wouldashouldacoulda even followed me to ABC Records years later. At the height of the Haight-Ashbury era of love, peace, protests, psychedelics, and flower children, our president Larry Newton sent me to San Francisco to find and sign up as many of that city's hot new rock groups that were causing so much upheaval and excitement throughout the record industry. This was during a transitional era in the music business when most record companies were clueless about where music was going and what would sell.

The late and still revered jazz critic Ralph J. Gleason took me all over the Bay Area in my quest to discover California gold (record stars, that is) and platinum, as did the excellent San Francisco photographer Jim Marshall and my friend Al "Jazzbo" Collins, the city's adopted son, favorite disc jockey, and municipal mascot. In fact it was Jazzbo who took me to hear local legends Big Brother and the Holding Company and their then unknown lead singer Janis Joplin.

The warehouse-instantly-transformed-into-a-sold-out-concert-venue (a staple of that current Bay Area rock culture) was a riot of people on line, on line to get on line, swarming at every barred entrance, standing on top of cars, selling T-shirts, drugs, liquids, scents, literature, playing guitars, etc. Jazzbo knocked on the door to the box office, announced himself, and was advised, "Sure you can come in. But, please, when we open the door, rush in as fast as possible."

The door opened, barely wide enough for us to squeeze in before it was quickly shut, and we were then immediately immersed knee-deep in money—a record executive fantasy come true! This was before credit cards and computerized ticket services, and people were buying tickets in such a frenzy that the box office ladies simply threw the cash over their shoulders, where it piled up on the floor behind them.

Another night Ralph Gleason took me to meet Steve Miller at a party so crowded the only place we could talk was seated on the floor under a grand piano. The conversation went amazingly well.

Steve finally said, "Gee, you're the most honest guy from a record company I ever met. I'd love to sign with ABC, but I need a $100,000 advance to build a small recording studio in my house." I told him not to sign with any other company as I would get back to him with an answer as soon as I could.

I heard and saw many great bands in San Francisco and rushed back to New York with my report to Newton and my enthusiastic recommendation that we immediately sign Big Brother with Janis Joplin and Steve Miller before another label got them.

Typically, when Larry Newton saw the probable six-figure advances and comparable recording costs, he said, "Forget it, absolutely not. What else did you find?" I then half-heartedly told him about a group of beatnik characters named Salvation who offered to make an album for $5000. Newton instructed me to sign and record them with the astute reasoning: "They're the cheapest, they're the best."

Columbia got Big Brother with Janis Joplin as well as Janis Joplin separately, and the Steve Miller Band signed with Capitol where he had three #1 singles. We made two rather good albums with Salvation, but Salvation turned out to be just a name, and not a prophesy.

III

Then there was the time that I became the object of corporate affection, instead of being a participant in a bidding war with another record company to sign an artist.

I had been with Coral, the "black sheep" division of Decca Records, for about nine years, had more than my share of hits, and had just been refused a $50 raise by Milton Rackmil, the president of Decca, when Mannie Sachs called me.

A power and legend at Columbia Records, where he produced his friend Frank Sinatra as well as Tony Bennett, Perry Como, and a galaxy of major vocalists, Sachs had recently joined RCA to form his own record company. The new label was to be called "X," and he wanted to hire me as his A & R chief for $50 more each week than my Decca/Coral per annum. Disappointed and bitter at my present bosses and their excuses —("It's the division budget," the perdurable classic "I wish it were up to me, but it's out of my hands," and the all-time favorite "There is an executive salary freeze from upstairs")—and eager to work for an industry giant I greatly respected, I promptly said, "Yes, I'll take the job. This sounds good."

Soon after, and like a kid who decides to leave home for the first time, I became conflicted with second thoughts about leaving my "family." I couldn't bring myself to tell Rackmil I was quitting, even though I had promised Mannie Sachs I would report to RCA on a Monday in two weeks.

A week went by, and suffused with emotions of guilt and cowardice, I hadn't done anything. Then I got a call summoning me to the executive offices for a meeting with Rackmil and Leonard Schneider, his vice president.

When I got off the elevator at the Decca executive floor, I ran into Mike Connor, the company head of promotion and publicity, who greeted me with, "Hiya, Bob, what are doing here?" To my reply that "I have a meeting with Rack and Schneider," Mike said, "You mean about the *Billboard* story?" I got very concerned. "What *Billboard* story?" He laughed, "You'll find out."

Rackmil got right to the point. "Bob, Len and I have been thinking about your services to the company, and we've decided to give you your $50 raise." To which I, in grateful, relieved idiocy gushed, "Gee, thanks," and, not to invite any further discussion, walked out.

What I did not stick around to learn was that *Billboard* had telephoned them to confirm the story I was leaving to become head of A & R at RCA's new X label, and to ask who Coral had chosen to replace me. Rackmil then called Mannie Sachs and told him, "You can't steal our top Coral A & R executive like that," and Mannie, who had survived enough intermural industry wars without needing another one, immediately agreed.

With ignorance of this presidential exchange added to the increased guilt and timorousness this new crisis created, I finally called Sachs. Before I could finish saying, "Mannie, I have to tell you something," he said, "I know, I know. You aren't coming to work." I was amazed. "How do you know?" "Rackmil called me. They're not going to let you go."

I stayed at Decca for a short while longer.

10 Pay Misty for Me

ART Blakey, the immortal jazz drummer, always fondly told the story of how he was introduced to his instrument.

A Pittsburgh native, and son of a shoe store proprietor (as youths Art and a childhood chum who grew up to be the most recorded bassist in jazz history, Ray Brown, both shined shoes in Papa Blakey's shop), the younger Blakey was originally a pianist, who, in his teens, had gained the requisite journeyman proficiency to provide musical background in those "colored section" bars where health codes and child labor laws allowed him to perform for indifferent customers.

One night, after several weeks of performing at the same establishment and beginning to believe he had found steady employment, Art was told by the owner to let this little neighborhood black kid named Erroll Garner play a few tunes. Within minutes, a dump became a joyous cathedral; the usually inattentive and boisterous customers were whispering offers to buy each other drinks; and an impressed owner knew that lady fortune had just smiled ear to ear toward him.

When young Mr. Blakey attempted to reclaim his place on the

piano stool, his boss, who wore a loud sharkskin suit, spoke coarsely, and carried a pearl-handled lump in his jacket, indicated he was not amenable to debates. He advised, "If you want a job here, go play those," and pointed to an ancient set of drums in the corner.

My first meeting with Erroll Garner was, to say the very least, a much happier encounter.

It was March of 1945, and I was trying to make my first label, Signature Records, into a business when a friend, Billy Moore, the great and respected Jimmie Lunceford musical arranger, telephoned to tell me I should immediately rush to his office and hear this young pianist from Pittsburgh who must be recorded. ("Man, you won't believe your ears!")

A few minutes after I arrived, shook hands, and heard this impish, smiling, and energetic musical genius play, I was arranging the studio time that we used to record two original Garner compositions "Loot To Boot" and "Gaslight," along with the standards "Yesterdays" and "Sweet Lorraine" a couple of days later.

Erroll was my first "discovery"; Signature Records now had four classic jazz sides in its catalog; and so began one of the most treasured friendships of my life.

Someone who could instantly captivate any audience he confronted, whether it was 20,000 people in a stadium or a booking agent across an office desk, Erroll Garner was, as I am still, essentially a loner.

Our friendship deepened as Erroll and I ran into each other at those famed New York City jazz clubs, mainly on 52nd Street, that thrived during that middle-1940s glorious Golden Age, when any evening we could walk into one of many nightspots to socialize with and hear our mutual friends and heroes Charlie Parker, Art Tatum, Coleman Hawkins, Billie Holiday, Ben Webster, and many others perform with their all-star groups. Erroll was always alone— I cannot recall ever seeing him arrive or leave with anyone—and we sort of gravitated to each other.

He also enjoyed having a daily lunch—usually alone—at the China Pavilion on Seventh Avenue and 57th Street, and let me know I was always welcome to join him, an invitation I accepted with increasing frequency.

As the years continued, and Erroll became a beloved jazz star while I landed at Decca Records to start a productive decade as chief of the Coral Records division; our friendship remained strong and as constant as our careers and schedules permitted.

I began hearing on the radio a gorgeous melody that I loved from the first Garner solo recital on Mercury Records that I soon learned was a new ballad titled "Misty" he had written. When I next saw Erroll and told him of my high regard for his new composition, he was very pleased and asked if the song could, in any way, be helpful to me. After my half-joking reply that I would love to be the publisher of "Misty," his response was a magnificent surprise: "Bob, you got it! I know I can trust you, and you will do everything you can to develop the song." Of course I was deeply touched and grateful for my friend's very generous gift of both this copyright and his unconditional confidence in me.

Again, the music business operated much differently in those days. Artists and repertoire people like myself did not make the big five- and six-figure salaries prevalent today, and it was assumed, expected, and necessary that record executives on my level would supplement their income from other related sources. At that time, the ownership of a publishing company and/or writing songs was considered entirely appropriate and without any ethical conflicts. Everyone in what was still a rather small professional music community knew each other and what everyone was doing. And as this was many years before abuses such as payola, drug transactions, and multiple sets of books became commonly practiced as a result of a rapidly expanding industry and the coincident reduction of honest standards, these "side businesses" were unquestionably proper. Besides, the record consumers of that era were far more discriminating than those of succeeding and oversaturated generations, and the corporate politics were also much simpler: you still had to make good records that sold to keep your job.

My publishing company was named Vernon Music (the town of Mount Vernon, New York, was near my home then), and by producing recordings of my copyrights and songs whenever there was an opportunity at either Coral or Decca, Vernon had become an enormously profitable enterprise. In rapid succession, I was able to make at least a dozen instrumentals of "Misty" for various situations, which both vindicated Erroll Garner's faith in me, and kept my creditors as silent rather than active partners.

The realization that I was utterly incapable of presiding over the daily operations of a burgeoning business—I was a creative type, not a bureaucrat—persuaded me to reunite with Chester Conn, my childhood neighbor from Forest Hills whose powerhouse publishing company, Bregman, Vocco & Conn, would administrate Vernon Music for me. No money was exchanged to complete the affiliation since I knew I could completely trust Chester Conn, among the most respected and distinguished veterans of the music industry. Additionally, one of my dearest lifelong friends, Frank Military, was employed at the firm.

It was Frank, in truth, who suggested that the lyricist Johnny Burke—who wrote "Pennies from Heaven," "Imagination," "Like Someone in Love," "What's New?," the Crosby-Hope "road picture" classics, the Oscar-winning "Swinging on a Star," and numerous other standard songs—add words to the melody of "Misty." Within weeks, Frank got the first recording of the music and lyrics with Sarah Vaughan on Mercury, followed by a huge success by Johnny Mathis on Columbia, and soon after, by everyone else. Perhaps I "found" the song, but it was Frank Military who found its lyricist and the initial recordings that made "Misty" an enduring American popular song standard as well as a perennially remunerative copyright.

Several years later, I had a son, Bobby Jr., a house in New Rochelle, and a particularly nasty divorce that totaled me! I am still not entirely sure what machinations culminated in the eventual settlement, but I ended up with temporary custody of my five-year-old son and the house, and my ex-wife got custody of substantially every dollar I had, could get, or would have from any foreseeable future earnings.

Suddenly faced with the full financial responsibilities of child-rearing, the upkeep of a large suburban house, and a live-in nurse for my son while I worked in Manhattan, I started to sell off everything I could. Furniture, paintings, chandeliers, cars, silverware; every week I would have to sell something else for cash to meet the overhead that kept pulling me under. Finally, all that was left in the house was a refrigerator, three rooms (mine, Bobby's, and the nurse's), and two beds (Bobby's and the nurse's).

My only other retail asset was my Vernon Music stock, which I then sold off to Lawrence Welk, Sam Lutz, who was his personal manager, and Lawrence's musical director, George Cates. I ended up with 0 percent of Vernon and its "Misty" copyright, and my about-to-be millionaire friends in the music business garnered great wealth from my marital troubles.

People say if you have your friends and good health, money is not so important. I've always been healthy, and enjoyed many cherished friendships.

But I can't help thinking . . .

11 The Bad and the Beautiful

I

WHEN Teresa Brewer and I celebrated our twentieth wedding anniversary in 1992, it was another reminder that these many years in the music business were always more beautiful than bad.

Through the decades I have been in the record industry, most people, professionals and civilians alike, have associated the word "bad" with a word of contemporary vintage that was originated within the music business and is now indelibly attached to it: payola—the payment of cash or other bestowed inducements by record promoters and executives to radio disc jockeys for the essential repeated airplay that generates record sales.

In these days of pervasive mega-conglomerate ownerships of all communication media, radio included, payola is virtually obsolete. Disc jockeys are now basically hired voices, and programming decisions are the purview of music and program directors whose corporate survival is based solely on station audience ratings. In turn, the ratings are dependent on the mix of recordings these directors air, and the abundant perquisites and six- and seven-figure annual salaries they earn far exceed any "gifts" they could receive from industry promoters.

Similarly, when I was at the height of my "pop" power, hit records resulted from the most appealing combination of song material and performer, and a producer who blends those elements. No amount of money has ever been able to "buy" a hit, then or now, if the record wasn't well made to begin with.

Of course, money, gifts, and temptations were always around, and before the era of confused transition, during which rock permanently replaced pop as the predominant commercial commodity of the music business and the record industry suddenly expanded to unparalleled proportions, cash and other benefits were simply professional courtesies and the normal costs of doing business.

The profession I was a part of before the rock era completely transformed the music industry was a small community of record producers, music publishers, disk jockeys and artists who had known each other for years. When I first became artists and repertoire director for Coral Records and sought advice from Milt Gabler, my venerable mentor whom I was replacing so he could move up and head the Decca Records creative department, all he said was, "At Christmas, you better rent a station wagon because of all the gifts you will get."

And even before that holiday, cases of liquor, kitchen appliances, luggage, etc., would regularly appear, and, when sleighbells rang, BMI sent over a grandfather clock and someone else gave me a washing machine the day after a bedroom set had arrived. It was still an era of innocence and probity when music publishers could give A & R executives checks "in appreciation" throughout the year ("appreciation" all of us would declare on our tax returns), and producers were actually encouraged to write and publish the songs they recorded.

Similar to government lobbyists or, in fact, what is customary in every other industry, these expressions of gratitude among longtime and collegial professional associates were merely considered good manners with no strings attached. And since these normal displays of generosity never directly influenced what we recorded or would be played on the radio, no one ever thought industry ethics or anyone's integrity was compromised. We still could keep our jobs only by producing records that would get on the air and sell in the stores by virtue of the fact they were good records, and not because cash had been transacted.

Then again, the checks I received away from the Decca/Coral Corporation were never for inordinately large amounts. My salary was at the level comparable with that of all the other A & R men at the time, for I guess our bosses knew we would pay them to be employed in the record business if we had to. Thus whatever monies came our way were always a matter of survival rather than influence. Also, I cannot recall any personal involvement with "payola," as it became commonly known and prevalent.

After a while, it was impossible not to be aware this had become the new way of doing business in a dramatically enlarged industry. A good friend of mine at the time was the colossally powerful disk jockey Alan Freed, who, later on, engineered the use of payola to such uncontrolled excesses of avarice he thoroughly destroyed his immense career and then his life in the process.

Alan was always kind and accessible to me, even though he was such a popular radio personality in New York City—then as now *the* major media market in the country. Unless he played your record on his show it would not be a hit, but he never asked me for payment and was always pleased to give my pop singles the vital exposure on his show that would guarantee their success.

Perhaps he didn't need my money, since there was invariably a long line of people with envelopes of cash waiting outside his studio and office whenever I visited him at either location. Without ever any monetary tribute or retainer from me, I was always assured of a warm and immediate welcome.

I also remember a visit to his stately home and grounds on the Stamford, Connecticut, beach front that overlooked Long Island Sound. Alan had this lavish estate with a private beach (dotted with portable urinals—the ultimate of conspicuous consumption in those days), and an olympic-size swimming pool so near the water that ocean waves would constantly splash into it.

When I asked him how, even with his tremendous success, he could afford the upkeep of such spacious property, his memorable reply was, "Atlantic Records does the lawn, Jubilee Records cuts the shrubs, Scepter Records cleans the pool. Everything is taken care of. I'm covered."

Admittedly, I also produced many records with Alan Freed and his fantastic tenor saxophonist and musical director, King Curtis, and I never learned whether those recordings substituted for the seemingly requisite payments I was exempted from. I do know that Alan asked me to be his producer, and as we were good friends and I adored King Curtis's consummate musicianship from his days with Lionel Hampton and Nat King Cole (whenever he was available, I would also hire Curtis for my sessions with the McGuire Sisters), it would have been pointless to refuse the invitation. I reasoned it would be foolish, if not disloyal to my bosses, to step aside for another producer and the label that employed him so that they would profit from the massively advantageous airplay and personal appearance momentum the Freed/Curtis sides would inevitably attain. (Aside from the radio exposure, the multi-star Alan Freed Concerts were, of course, as major an entertainment event as was imaginable anywhere.) Besides, and again, they were good records I will always be proud that I produced.

As previously recounted, I never participated in the less savory practices of the music business that began to burgeon during my "pop" period. Not that I might have influenced any of these increasingly prevalent activities, but it was difficult to be unaware that sometimes even my records were part of those iniquitous pursuits to which corrupt music promoters throughout the country had become devoted.

In truth, the only time I was ever personally entangled in the music business underside was a controversy that involved my employer, Decca/Coral Records, one of our main competitors, the Capitol Records pop music titans Les Paul and Mary Ford, and the now-deceased executive who headed their music publishing companies.

This began with a summons to the offices of Decca Corporation treasurer Isabel Marx, who unceremoniously greeted me with the jolting news that my employment was to be immediately terminated unless I could somehow explain an action that had gravely upset my bosses. I was staggered, since I couldn't imagine what had provoked this crisis. (My temper tantrums and an innately rebellious nature had been routinely excused in light of the fact that I was creating more hit singles than any other staff producer.)

I learned that the attorneys representing Les Paul and Mary Ford (at the time Mrs. Les Paul) had advised the corporate lords at Decca that for almost a full year the Paul/Ford music publishing company had paid Bob Thiele $75 a week as he had demanded to record their copyrights. Since none of the recordings had materialized, they now intended to sue Decca and myself for fraud in a civil lawsuit while they simultaneously would provide the District Attorney with proofs for a criminal indictment. Further, the lawyers had produced an incriminating and imposing stack of canceled checks made out to me with my endorsement.

Not having had a hint anything like this was occurring, and certainly never agreeing to this sort of deception, I naturally asked to see the evidence. Once confronted with this proof, I said, "Isabel, if you had taken the trouble to compare my signature on these checks to all of my Coral Records salary checks, you can see that the two endorsements are so different, that the same person could not possibly have signed both sets," a fact even she could not deny.

Subsequent investigation soon revealed that the real culprit was the professional manager of the Paul/Ford publishing firms. In this instance, it was discovered their employee had convinced Les and Mary of the complete fiction that he was "very close" to Bob Thiele, the hottest young producer in the business, and for the meager $75 a week Thiele insisted be personally delivered to him (a bargain fee as a favor from one "old pal" to another) Thiele would arrange for numerous Paul/Ford-controlled copyrights to be included on his forthcoming Coral and Decca productions. The checks were then endorsed with my name and cashed, the perpetrator inexplicably and destructively thinking his deceit would never be exposed.

Of course the swindle rapidly unraveled once it was shown my signature had been expropriated, and I believe a deal was arranged with the D.A. and Les Paul and Mary for the executive to make the full restitution that could not save his career but kept him out of prison.

For me, however, the worst "bad" of all in the music business are the jazz critics who seem to increase to such an extent that they now outnumber the musicians. Most of these self-anointed "experts" probably could not survive in any profession, and they certainly do not understand how music is performed or created. From years of horrified observation, I can only conclude that the main priority of most critics is to find a place—anyplace—to fit in, and the jazz scene tolerates these pretenders more than most such groups would. Consequently, the music continues to contend with critics whose aesthetic standards are abysmal in their ignorance.

They simply want to use the validation of the printed word to justify their professional and personal inadequacies.

To be even-handed, writers such as Leonard Feather, Nat Hentoff, Dan Morgenstern, and Stanley Dance have provided those touchstones of objectivity, honesty, and knowledge that are regrettably more exception than rule in the jazz world.

It's those others with their pretentious pronouncements and pernicious arrogance that annoy me. Most of the time, after reading a review, and knowing I've heard more music, worked with more important musicians, and produced many more valuable records than the vast majority of these so-called authorities ever will or could, I continue to wonder, "Who are these people anyway?"

II

Beauty also has numerous definitions, and I remain thankful my career has allowed me to experience beauty in a number of ways.

The associations with so many exceptional jazz artists would top any list. It would include musicians who have shared their friendship and immortal talents with me while I was permitted and honored to document some of their most definitive work for the generations that would follow them: Coleman Hawkins, Lester Young, Louis Armstrong, Erroll Garner, Shelly Manne, John Coltrane, Dizzy Gillespie, and my eternal pal, Duke Ellington, to name a few.

I must also mention Steve Allen, a one-time business partner as well as a longtime friend, supporter, confidant, and collaborator whose contagious gifts of humor and talents are manifold. Even more, he has an extraordinary ability to bring out the best in others, to explore every artistic possibility, and to maintain the highest professional standards.

While in truth jazz will always be the primary focus of my music business activities, I have had the added pleasure of producing records with many film celebrities—from Don Ameche, Rhonda Fleming, Jane Russell, and Jackie Cooper to an effervescent dynamo named Debbie Reynolds, who had her only #1 record, that indestructible paean to pubescence, "Tammy," with me. Despite the stereotypes that are usually ascribed to movie stars— obsessed with their looks, their overwhelming egos, their mindless remarks and actions—I have found them to be among the hardest-working professionals I ever dealt with: unfailingly respectful of the music and their obligation to perform it faithfully.

Of all the movie names I recorded, it was the gifted actor George Segal who provided some of my fondest memories. I first saw him on the Johnny Carson *Tonight Show* and, along with the rest of the nation, was surprised that he played a musical instrument, the banjo, in a more than passable manner, and unashamedly sang song classics from the earlier jazz eras with an infectious passion. I found this musical talent made him seem an even more compelling star than I had considered before. Besides, he was instinctively musical, swung like mad, and I wanted to make a jazz album with him for Flying Dutchman, the RCA subsidiary label I owned at the time.

A couple of more surprises followed. I learned that George Segal was represented by none other than the storied former Swing era band leader (and my once idol) Shep Fields, now a power at the William Morris Agency. Further, Maestro Fields, thoroughly amused that anyone with a reputation in the music industry would think to make a record album with his illustrious client, then quoted me a price of $50,000—an exorbitant fee, movie star or not, that would bankrupt my bankroll and give my accountant, if he had any heart at all, a coronary.

As any other obsessed record producer on a (by now) consecrated mission would have done, I promptly schemed and scammed every contact I knew in the Hollywood community to obtain Mr. Segal's private home telephone number. Soon after, I was treated to one of the more singular conversations of my life, and a result, gained a marvelous new friend and recording artist.

When the ebullient, famous voice answered his phone, I still have no idea what impelled me to say the words Jelly Roll Morton had recorded over forty years before, but the first sentence George Segal ever heard me say was, "Hello, Central, give me Doctor Jazz!"

His response wasn't "Who is this?" or "How did you get this number?" Instead, "You know that tune? Whoever you are, we just became friends!"

I introduced myself and related my vision of a George Segal jazz album, a concept he had never thought of or, despite my recent exchanges with his agent, had heard about. Immediately with equal excitement and caution, he began peppering me with questions such as what songs did I have in mind, instrumentation, where would we record, when, etc.?

Increasingly persuaded but not yet entirely convinced, George then asked which musicians I would want to use. I mentioned some of the all-star jazz names I frequently recorded with, and when I told him the trombonist should be like Georg Brunis, a revered albeit arcane name from the Swing era's best years, the reaction was instantaneous.

"You know who Georg Brunis is—we *have* to make an album together," and in the next minute George Segal had agreed to fly to New York, stay at his mother's house, and record with me, all for an inconceivable $2500 fee!

The studio session that soon followed was a revelation. I had gathered such jazz greats as Hank Jones, Pee Wee Erwin, and Bucky Pizzarelli in the New York City RCA studios to accompany George, my wife Teresa Brewer—who had consented to make a special guest appearance on the album—and a young, soon-to-be-legendary unknown singer-pianist-composer named Harry Nilsson, whom George Segal had befriended and now insisted also be a participant. Segal, this screen idol whose face was known in every corner in the world, who had performed with the most glamorous women, appeared before royalty, and moved among the notable and wealthy, had been the earliest to arrive and could only be described as the happiest little kid who ever visited a toy factory or circus.

As soon as George, in that irrepressibly animated elation millions of television viewers were now familiar with, began to sing and play the Harry Von Tilzer chestnut "What You Goin' To Do When the Rent Comes 'Round (Rufus Rastus Johnson Brown)," the first song (I believe) we had chosen to record, the steely, suspicious jazz veterans, who had previously and bluntly voiced their serious skepticism to me about recording jazz with a movie star when I had hired them, were all at once smiling, nodding their heads to the rhythm and playing as well as I had ever heard. Suddenly, we were all giddy revelers at one of those all-too-few magical celebrations where everything catches fire, goes right, and everyone is invigorated to produce at and beyond their peak levels of creativity and accomplishment.

With his energetic jubilance, and complete lack of pretension,

George Segal was the catalyst for that most rare, marvelous occurrence—an occasion no one wants to end or let go of. Busy, fiercely
professional musicians were begging for another take because they
wanted to contribute even better performances, and then staying
after the microphones were turned off to hear every playback.
Harry Nilsson, now two hours late for a record session (and eventual career milestone) with John Lennon after our date, stayed
even later to perfect a four-measure piano introduction that everybody except him had thought could not be improved on.

One of the nicest, good-humored, and enthusiastically cooper-
ative people I would ever know in a half-century-long music business career, George Segal had made the record album of his life,
and given the rest of us an experience as unforgettable as it was
unexpected. I treasure his friendship and eagerness to share music
with us.

Then there are those superlative industry veterans "on the
other side of the desk," music men who have always cared enough
about the lifeblood of creativity to make the good things happen
for artists and producers and the audiences they embrace. Ken
Glancy, the former president of RCA Records, is a member of this
admirable brotherhood, as are Frank Military and Murray Deutch.
A score of years before the concept became an industry cliché,
when my corporate superiors very brusquely rejected my idea for
record labels that would have pictures of the artists and individual
graphics and colors appropriate to the specific music ("too expensive," "nobody looks at labels," "impractical," "won't sell records,"
"ARE YOU FUCKING CRAZY!!!," etc.), I can recall it was Frank
and Murray who gave me the much-needed reassurance I was not
a candidate for the asylum and someday I would be credited for a
brilliant vision.

My list of corporate heroes should also include Bruce Lund-
vall, currently president of the EMI-acquired Blue Note Records,
and arguably one of the best friends jazz has ever had in the record
business.

And I should single out such legendary music publishers as
Abe Olman, Bobby Mellin, and Freddy Bienstock, who presently
heads the largest independent publishing conglomerate in the music world and years ago was Colonel Tom Parker's principal
adviser in choosing writers and material for Elvis Presley, the
Colonel's most renowned client. More recently, Freddy is a val-

iantly loyal and patient co-owner, advocate, ally, defender, con-frere, and proselytizer with Red Baron Records, my current record label that Sony Music Entertainment Inc. (formally CBS Records) distributes.

Finally, at Sony, are two corporate saints and industry para-gons who more than generously share their fraternity, encourage-ment, immense auspices, and toleration with me: Mel Ilberman, the chief operating officer of the International Division and a real friend since I had my Flying Dutchman company of many decades past at RCA Records, and Paul Smith, president of Sony Music Distribution and an indefatigably perennial pal I have known for even more years.

Not last nor least is Teresa Brewer, my brainy, beautiful, and abundantly talented bride of nearly twenty-five years who is my fourth wife, first lady, and lifetime best friend. When Teresa agreed to marry me after a courtship as brief as our acquaintance had been lengthy, I was certainly no expert on that hallowed institu-tion, but an unchallenged authority on matrimonial disaster, as three previous and mercifully short-lived experiments in wed-locked, gridlocked, and locked-out-of-my-house spousal terror would corroborate.

My first marital misadventure came about when I was a much-too-young 21-year-old, and it lasted only ten months. Fiasco num-ber two endured for five years while I was wedded to vocalist Jane Harvey, the mother of my only son and other best friend, Bob Jr.; and my third marriage lasted only a year.

(I am, of course, extremely proud of my son, Bobby Jr. I remember our forced bachelorhood together after his mother and I were divorced when he was the only boy in the neighborhood to sing and perform Elvis Presley songs with his small guitar while listening to John Coltrane records at home. Both of these musical influences as well as many others proved to be most beneficial as Bob Jr. is now a much-in-demand songwiter, record producer, and session guitarist who works with such pop music icons as Bonnie Raitt, Aaron Neville and his brothers, and the great Ray Charles.)

With Teresa, I remember being in love with this warmly charming, exquisite woman from the time of our initial meeting in the early 1950s when I was a young producer at Coral Records and assigned to record her. A Major Bowes' Amateur Hour winner and show business veteran, Teresa was already a major star. "Music,

Music, Music (Put Another Nickel In)," one of the mammoth hit singles of the decade, had established her as among the prevailing pop vocalists of her era. It was my responsibility to keep the Brewer hits coming; a task made nearly effortless and easefully pleasurable by Teresa's incomparable adaptability, professionalism, generous personality, and unerring musical instincts.

Starting in 1952, when Teresa Brewer and I began our nine-year, unbroken string of hit single successes (which included such Top 10s as "Till I Waltz Again with You," "Ricochet Romance," "Let Me Go, Lover!," "A Tear Fell," "A Sweet Old Fashioned Girl," and "You Send Me"), I was intensely in love with her, but could not say a word about my feelings. We were both married, and Teresa was a rigorously proper Catholic lady with four daughters and a large, decent family for whom any hint of scandal or the thought of divorce would have been profoundly abhorrent.

The first #1 hit single that we recorded together was "Till I Waltz Again with You," a song Teresa first heard in the elevator of the Brill Building, the now-mythical music business landmark in New York City, when the writer, Sidney Prosen, auditioned it for her between floors. It was originally conceived as a country song and, the title notwithstanding, wasn't even a waltz. Teresa loved the melody and words, but as a country lament the song never worked. When we transformed it into a pop tune, "Till I Waltz Again with You" whirled to the top of the hit parade, where it was #1 for seven weeks of the half-year it stayed in the "Hot 100." Even after all this time and all the material Teresa has since performed, it remains her most favorite song that she has recorded.

Our following huge success the next year, "Ricochet Romance," could easily have been a huge calamity. When we recorded the song, overdubbing was still an uncommon and impractical luxury as everything—vocals, the orchestra, instrumental solos—had to be right coexistently, and any additional repairs or improvements would have an enormously expensive impact on the production budget and, if the song was a hit, on the royalties— the eventual earned profits—received by the artist.

Well prepared as usual and with her typically buoyant proficiency, Teresa sang the tune in the few perfect takes we needed and went home. It was after the recording, and during the ensuing days as I listened to the results of the session, that I became increasingly unhappy with how Teresa's vocal was balanced against the orchestra. Soon after, and despite the extreme costs involved, I began to campaign for another attempt at this song that had "monster hit" written all over it. I finally persuaded Teresa to participate in a second session that would guarantee a commercial triumph which would more than abundantly compensate for any deductions in her royalty account.

Trusting, not entirely convinced, but masterful pro that she was, Teresa amiably and expertly provided the performances I asked for at a subsequently scheduled record date. When I reviewed these later efforts, I then realized it was the original, earlier sound that would give us the hit.

Without telling her, I then released the first recording we did. When "Ricochet Romance" arrived at the stores and radio stations, and Teresa figured out what I had decided to do, she was, in rapid order, puzzled, disillusioned ("I thought you guys knew what you were doing!"), and apprehensively suspicious, although having a Top 10 single that stayed on the charts for another half-year quickly overcame her objections.

"Let Me Go, Lover!" was the next smash for us. During that era, if a similar or competitive artist had a hit, or even perhaps *might* have a hit, with a certain song, there would be a frantic scramble to release "cover versions" in order to capitalize on the popularity of the original. "Let Me Go, Lover!" was the theme song from a highly rated and publicized live television drama on the CBS network, the now-historic November 15, 1954, *Studio One* telecast that was, ironically, a fictional exposé of the music industry. An unknown Mitch Miller protégé named Joan Weber had recorded the soundtrack version, and every major label producer had their star female vocalist in the studio to cover the tune within hours after the program went off the air.

I recorded Teresa at 2 a.m. the next morning so we could rush out a single, and both Patti Page on Mercury and Sunny Gale with RCA also hurried out their versions of "Let Me Go, Lover!" to challenge the Joan Weber single, which Columbia Records had issued and presciently had stocked in the record stores weeks before the *Studio One* show would be broadcast. As it turned out, all four recordings went Top 20 within the same month, each selling a million copies (Teresa's went to #6 and continued on the charts for 12 weeks). Interestingly, only Joan Weber's single alone captured the #1 slot with what was to be the only hit record she would ever have before she completely disappeared from the music business.

Three of the other Top 20 Teresa Brewer singles after "Let Me Go, Lover!" were also covers, in these instances, of "rhythm and blues" or, as they were then called, "race records: songs by black artists who did not have the same access to the much larger and richer white "pop" market. Once these songs became proven hits in the black arena, they were unreservedly appropriated by producers such as myself for their major white acts. Teresa was always uncomfortable and guilt-stricken with this preponderantly common industry practice, and I was never able to change her feeling that this was unfair and exploitive, despite my continual explanations that this was the record business, and that if we didn't cover black artists, everyone else would still engage in the practice, and we would lose the hits we could have had.

The songs were "A Tear Fell"—another of Teresa's personal favorites—which hit #5 and charted for six months, the great Ivory Joe Hunter song "Empty Arms" (that one remained in the Top 100 for 17 weeks peaking at #13), and, in the next year, 1957, the Sam Cooke classic "You Send Me," which crested at #8 for Teresa during its three-month residency on the hit parade.

Also, half a year after "A Tear Fell," I produced "A Sweet Old Fashioned Girl" with Teresa, a Top 10 single with a 20-week chart life, and among the very first records that deliberately and effectively bridged the old traditional pop sound with the rapidly spreading new rock 'n' roll style that was then ascending to its irrevocable primacy. I liked the song for Teresa because it could accommodate both the old and new pop music audiences so well. Moreover, "A Sweet Old Fashioned Girl" actually sounded like a dozen other tunes and, as with many other Brewer hits that had familiar melodic strains to them—such as "Music, Music, Music," which liberally borrowed from the second Hungarian Rhapsody by Franz Liszt, and "Choo'n Gum," with its obvious quote from Gershwin's *American in Paris*—the song fit the oft-demonstrated Teresa Brewer hit record formula perfectly.

I didn't then comprehend it would be necessary for another decade to pass before Teresa and I would be free to act on our personal feelings toward each other. After her husband had died, we began to date, and soon decided to marry. Incredibly, her mother Helen seemed to know this would happen months, if not years before we did. Teresa had not made the slightest mention of our plans or even that we were dating to anyone in her family, and I had not seen or spoken to her mom in many years. To be sure, we were thoroughly astounded when Teresa telephoned her mother back home to announce she was in love and about to be married, and Helen promptly said, "I'll bet it's that nice Bob Thiele you used to record with!"

All these exceptional years later, Teresa and I have traveled everywhere, and thanks to her phenomenal versatility, recorded jazz, blues, Hawaiian, pop, country, American Songbook standards, rock, ragtime, show tunes, and every other conceivable popular musical idiom with everyone from Duke Ellington and Count Basie to Conway Twitty, Mickey Mantle, and Peter Frampton.

Teresa continues to be the most compassionate and supportive alter-ego, partner, helpmate, and inspirational presence I have ever known in my lifetime, and after 25 years of an immeasurably wondrous marriage that has enabled us to share every possible professional and personal joy, we're still on our honeymoon.

12 The Woodpile

I

THESE days, it is essentially money, and ancillaries such as production budget overrides, sales escalators, percentages, ownership options, tour support, and advertising budgets, that determines the decision of an important artist to sign with a record company. A previous occasion when Lawrence Welk shifted his label affiliation is an example of how times have really changed.

I joined the California-based Dot Records label after I had worked for Coral, a Decca Records subsidiary where Lawrence Welk was one of their most valuable artists and where we had become good friends. When my tenure at Dot began, my boss, Randy Wood, summoned me. Wood, a label owner/president whose persona combined a wholesome rural naïveté with a steely metropolitan aggressiveness, gave me my first assignment: "Boy, if you could only get your friend Lawrence Welk to leave Coral and come with us, it would be great for our company." (Subsequently, I was also asked to bring other Coral stars as Debbie Reynolds and

Steve Allen to the label, where they both enjoyed considerable success and scored with two hit singles each.)

I was among the very few who knew Lawrence was not entirely happy at Coral Records, and, after many clandestine meetings and conversations, he finally agreed to consider a move. Welk also advised me that it was necessary that he have dinner at Randy Wood's Beverly Hills home to get to know him, his family, and the "team" at Dot Records before he could make a final decision.

An "intimate" (Hollywood style) party was quickly organized. I was to secure the requisite limousine fleet and also bring, as extra influence, Louella Parsons, the most powerful entertainment/gossip columnist at the time, and her boyfriend, Jimmy McHugh, revered as the songwriter of such standards as "On the Sunny Side of the Street," "I Can't Give You Anything But Love," and "Exactly Like You." Aside from each other, Louella and Jimmy also shared a fondness for liquor. After we were on our way to the party in the limo, Louella demanded we return to her house because they had forgotten to bring along a bottle of Scotch. The bottle was not a gift—both Randy Wood and Welk were famed and vocal abstainers who obsessively avoided gatherings, if not postal zones, where alcohol could be consumed—but was needed so she and McHugh could drain the contents during what promised to be a long and boring evening for them.

As befitted the occasion, the dinner table was a lavish display, and all of us—Randy Wood with his wife and children, Louella and Jimmy, myself and some other Dot Records executives and artists—sat down with Lawrence Welk to enjoy a splendid meal. Servants appeared, and, as the appetizer was presented, Randy asked that we pause for grace to be offered by one of his children.

I was at one end of the table and Randy at the other, and, while his youngest daughter spoke the devotions, everyone respectfully bowed their heads and closed their eyes (and I am sure that Louella and Jimmy immediately took advantage of the moment to have another drink). I happened to glance up at the same time as Randy Wood and was surprised to see that he winked at me—a most uncharacteristic action for him at such a personally solemn moment.

Later he explained, "Bob, I think that cinched it. Once I saw how moved Lawrence was when my little girl said grace, I knew he was going to sign with Dot Records."

And he was right! That simple instant persuaded Lawrence Welk more than the impressive surroundings, cuisine, guests, or inducements ever would. A major artist signing could never happen like that today, or ever again.

II

Dot Records was established after a series of consecutive hit singles by Pat Boone (of songs Little Richard originally recorded) that were produced by Randy Wood, the owner and president of the Tennessee-based label. Wood began romancing me to head his artists and repertoire department (and become his East Coast vice president) while I was still with Coral/Decca and growing restless. Two years before I actually accepted his invitation, Wood had generously offered me 10 percent of the stock in the company to work for him. He also told me he would move the company from Gallatin, Tennessee, to Hollywood, and sell it to a major picture corporation. I thought he was off his rock and turned him down. Some 24 months later, he had sold the now relocated Los Angeles label to Paramount Pictures, and I (sans any of the significantly higher priced current stock) was his newest executive acquisition.

During what was to be an abruptly abbreviated tenure at Dot Records, I had not only persuaded Lawrence Welk, Debbie Reynolds, and Steve Allen to leave Coral Records and join the label, but I had also created a critically acclaimed, steady-selling jazz catalog. Sadly, the relationship between Randy Wood and myself was plagued from the start. I was a music man; he wasn't. For me, music was a momentous force that made the world a better place. To Randy, who was an uncommonly impressive businessman in most respects, music was just another product to sell.

Randy Wood was also a monomaniacal moralist, unrelentingly preoccupied with his over-inflated edicts for global salvation. It was because he became so offended by my (and Steve Allen's) Jack Kerouac album, as well as my refusal to agree with his distended moral dictum that *Poetry for the Beat Generation* was a "pornographic outrage," that at the same time he pulled the Kerouac album off the market, he asked me to leave the company.

Further, I was losing my friendship with Lawrence Welk, who, like his new best friend in all of show business, Randy Wood, didn't drink, smoke, or swear. The more convivial they were, the more my relationship with Welk dwindled because I (and many times with great exuberance) drank, smoked, and cursed.

Wood had moreover become consumed with paranoia by my developing rapport with Pat Boone, the only Dot artist he produced and one he felt particularly close to philosophically and creatively. Boone was the star who began with him in Tennessee and whose hits had now made Wood an industry power and multi-millionaire.

Pat and I were in New York with a thirty-piece orchestra making an album of standards called *Stardust,* his first recording with a producer other than Randy Wood, who was telephoning the studio from his California office and home almost every hour. During one call, Pat got on the phone and said, "Oh, Randy, working with Bob is just terrific. The records are going great, and I really dig Bob." The next morning, Randy Wood was at the studio in New York, and all he could smell is that I'm going to steal Pat Boone from him when and if I leave the company.

I also suspected Wood's presidential ire had escalated as a result of a discreet (I thought) conversation I had with the actor, Dot recording artist, and my friend (I thought) Jackie Cooper, with whom I had made a couple of good albums at the label. We were at a party where I related an incident about a Mills Brothers album I had just completed for Dot and Randy Wood's strict admonition that I not use a picture of them for the cover because "We won't sell any records in the South if they see Negroes on our albums."

I had confided to Jackie that Randy Wood, in addition to his endlessly inordinate sermonizing, also seemed to be an unrepentant bigot. Although I could never prove that Cooper was the source, Wood found out I had expressed this opinion to one of his artists, and so our already deteriorating relationship worsened even more rapidly.

By this time, I was completely disgusted with Wood's constant, mindlessly fanatic preachifying (I could never get him to discuss music). As I now felt I was employed by a demented cleric rather than a music magnate, it was my eager inclination to immediately leave Dot Records as Randy Wood wished. My lawyer, however, counseled me to stay at the company since I had a contract with a year remaining at a high salary and benefits that would be forfeited if I resigned.

On advice from my attorney, I resolved to sit in the New York office for the next year to protect my legal rights and perhaps irritate Wood to the point he would fire me and I would still obtain the salary and perquisites specified in my contract.

It was not to be. I was unprepared for the stultifying boredom of inactivity during the endless days I was imprisoned at the office. I was not permitted to transact any current Dot business or future pursuits such as finding another job during the strictly-adhered-to eight hours I "belonged" to the Dot label and its owner. All I could do was sit at my desk, send out for lunch, and doodle on pieces of paper until the close of day.

Those scrap-paper scrawls quickly became my grasp on sanity as I took advantage of this one allowed exercise to formulate and structure the new record company, Hanover-Signature, Steve Allen and I planned to start when, concurrently, my "incarceration" and Steve's artist contract were scheduled to end.

Since I presumably could only "jot for Dot" in the office, I would tear up my notes each evening before I left. Unbeknownst to me, Mickey Adde, the executive who shared the New York office with me and was Randy Wood's bi-coastal flunky and East Coast promotion man (he had worked for Perry Como and Pat Boone before he came to Dot), had been instructed to empty out my garbage each evening and Scotch-tape together my discarded scribbles for shipment to the president's office in Los Angeles.

Within weeks, I was confronted with the irrefutable, taped-up (and coffee-, cigarette-, mustard-, and ketchup-stained) evidence that I had conspired to create a competing record label while in the employ of Dot Records. Nixon wasn't the first to be forced out of an office because of taping, and my lawyer advised me to settle out of court.

III

Along with many others, Steve Allen and I first became aware of Jack Kerouac through his poetry and, of course, his most important novel, *On the Road,* when the entire nation was transfixed with the artists and excesses of the "beat generation."

Steve and I had gone to hear Kerouac read his poetry at the Village Vanguard jazz club. Kerouac obviously loved jazz, and we were both deeply affected with his feeling for the music. The way he read his poetry, there was sort of a flow to it that made you think of jazz if you knew jazz. It made you think of a horn, as if Kerouac played saxophone.

For his first set, Kerouac came out alone, sat on a bar stool and started to read. It was a disaster. In that jazz club setting, and with no musical accompaniment, the audience couldn't understand what was happening. It really seemed as if Kerouac had shown up in the wrong place.

Steve Allen actually got the idea. He felt Kerouac needed to have some soft jazz in the background while he read his poetry, and Steve said he'd love to play piano behind Jack for the second set, an offer immediately and gratefully accepted by all concerned. When the dramatically more effective next set proved the correctness of Steve's instincts, we all began to realize the collaboration would make an interesting record for the Dot label, for which Steve was signed as an artist, and I was their East Coast A & R vice president.

On my authorization, we proceeded to record Jack Kerouac reading his verses, with Steve Allen playing background piano. The title of the album was *Poetry for the Beat Generation,* with liner notes by the venerable cultural essayist and critic Gilbert Millstein, and I promptly scheduled it for release.

Although the album quickly began to achieve an enthusiastic response from critics and industry insiders, I got a call from an apoplectically offended Randy Wood saying, "This record is *horrible*, it's almost pornographic. I wouldn't let my son listen to this record." And of course, I became indignant. "This record wasn't made for your son to listen to," I proclaimed. "This is contemporary poetry, and both Steve and I are sure it will be an important record release." (Which, in fact, it already was.)

Randy remained adamant, and immediately canceled all future pressings. It had also become increasingly evident to me that Randy Wood's autocratic obsessions with exaggerated morality were more important to him than making good records and music.

Shortly after that confrontation, both Steve and I left Randy Wood and Dot Records to form a company we called Hanover-Signature. Hanover, because I think Steve had accounts at the Manufacturers Hanover bank in New York, and Signature, because that was the name of my very first record company.

Bob Thiele, in a recent publicity photograph. *Photograph by Nick Sangiamo.*

Bob Thiele at age 5.

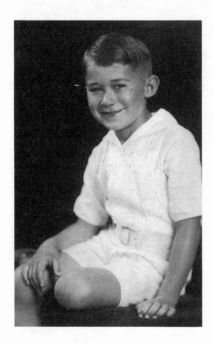

Bob Thiele at age 7.

Herbert Thiele, *circa* 1930. *Photograph by Bill Mark.*

At the International Hotel, Havana, *circa* 1935. *From left to right:* Irving Lundy, Bob Thiele's uncle and owner of Lundy's Restaurant; Henry Linker, Lundy's accountant; Bob Thiele, age 13; Athenaise and Herbert Thiele, Bob Thiele's parents.

Mickey Mantle with Teresa Brewer at the "I Love Mickey" recording session, 1956.

Bob Thiele with Mickey Mantle at the "I Love Mickey" recording session, 1956.

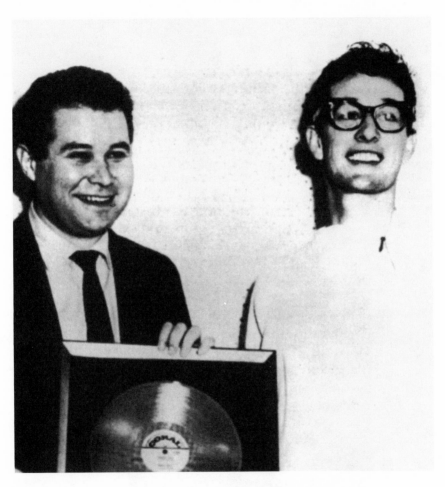

Bob Thiele and Buddy Holly at the 1957 Gold Record ceremony for
"Peggy Sue."

The McGuire Sisters, late 1950s. *From left to right:* Christine, Phyllis, Dorothy. *Courtesy of the Teresa Brewer Fan Club.*

Bob Thiele with Debbie Reynolds at the 1957 Gold Record ceremony for "Tammy."

Jackie Wilson in a rare 1960s publicity photograph. *Courtesy of Paul Tarnapol.*

Jack Kerouac with Steve Allen, 1959. *Courtesy of Steve Allen.*

John Coltrane and Duke Ellington, 1962. *Photograph by Bob Ghiraldini.*

John Coltrane and Bob Thiele, 1962. *Photograph by Bob Ghiraldini.*

Louis Armstrong on the sheet music cover for "What a Wonderful World." *Courtesy of Herald Square Music, Inc.*

Louis Armstrong and Bob Thiele, 1968.

Lawrence Welk, late 1960s. *Courtesy of the Lawrence Welk Show.*

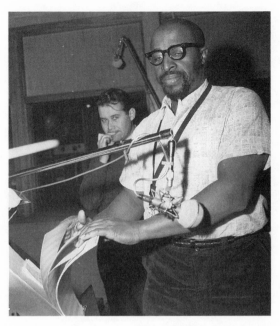

Bob Thiele with Yusef Lateef, 1970s. *Photograph by Chuck Stewart.*

Bob Thiele and Archie Shepp, 1970s. *Photograph by Chuck Stewart.*

Left to right: Oliver Nelson, Bob Thiele, Ron Carter, Hank Jones, 1970s. *Photograph by Chuck Stewart.*

Left to right: Muddy Waters, Bob Thiele, Otis Spann, 1970s. *Photograph by Chuck Stewart.*

Left to right: Count Basie, Bob Thiele, Duke Ellington, 1970s.

Left to right: Bernard "Pretty" Perdie, Earl "Fatha" Hines, Duke Ellington, Harry Carney, Teresa Brewer, Bob Thiele, 1973. *Photograph by K. Abé.*

Duke Ellington and Bob Thiele, 1973. *Photograph by K. Abé.*

Duke Ellington with Teresa Brewer, 1973. *Photograph by K. Abé.*

Steve Allen, 1984. *NBC photograph courtesy of Meadowlane Enterprises, Inc.*

Teresa Brewer.

Teresa Brewer, Wynton Marsalis, Bob Thiele, 1991. *Photograph by Hashimoto.*

Additionally, Kerouac became an acquaintance, and, as a follow-up album was sure to be both a major coup and sales event for our fledgling Hanover-Signature label, I asked Jack if he would be willing to record another reading of his poetry. He responded positively, and had only one demand: "I want to have my two favorite jazz musicians back me up this time."

I expected him to name traditional accompaniment instrumentalists such as a pianist and bassist. It happened, however, that the two favorite musicians—neither of whom he had met—were the legendary jazz tenor saxophone giants Zoot Sims and Al Cohn. My initial reaction was great concern. If Kerouac is reading poetry and there are two tenor saxes soloing against him, the clarity of the words might be lost over the music of the saxophones. But Jack insisted, and I went along. We then made a record which, despite my misgivings, turned out to be just fine and subsequently acclaimed as a classic union of heretofore disparate art forms.

When the recording was made, Zoot and Al were among the most in-demand New York City record-session musicians, ubiquitous on jazz dates, commercial jingles, soundtracks, mood music orchestras, and virtually every other musical situation that kept the city's record studios going from early morning to the next. The saxophonists approached this session as just another date—they performed, and as soon as they were done, they left (almost always to rush to their next recording job).

With the music finished and duly documented, Al and Zoot quickly packed their horns and, because of their inexorable schedules, left the studio without staying to say goodbye to Kerouac who, when the session ended, had inexplicably disappeared. Eventually, I found him in a distant corner of the studio crying.

I asked, "Jack, what's the matter, what happened?" It appeared that what had deeply upset Kerouac, who was as emotional as he was usually soft-spoken and pleasant, was that the jazz musicians he so idolized, identified with, and now felt were his friends as well as artistic colleagues had seemingly hurried from the studio without caring to hear the final takes of the session with him. (You just never know. You are happily satisfied that a record date has gone better than you ever dared to hope, and right in front of you, your—until that moment—rational and contented star has suddenly transmuted into an abysmally anguished, uncontrollable manic-depressive.)

In order to relax Kerouac out of his now rapidly intensifying despair, I decided he needed a beer—Jack loved to drink beer—at one of the many bars that lined Eighth Avenue in midtown Manhattan, the location of most New York City record studios, including the one we were in.

The nearest bar, like the others in the area, could only be described as a joint, where all customers would get a glass and a bottle, and pour their own beer. Jack had by this time become obstinately inconsolable and, after he drained each bottle, walked out of the bar, and defiantly threw the empty into the middle of Eighth Avenue. Fortunately, he didn't hit any cars or pedestrians. But he kept it up for quite a while. (It seems he may have hurled and smashed—as he very soon was—at least five bottles into the street.)

After the fifth bottle met the fate of its predecessors, I tried to use good judgment. "Jack, let me get you a cab. You go home, and I'll go home." But he wouldn't leave, and just kept drinking beer, breaking bottles in the middle of Eighth Avenue (to the hearty amusement of fellow customers and the bartender) and getting more drunkenly and desperately morose. I finally left, not knowing that would be the last time I would see Jack Kerouac.

13   Italian Roulette

THE relationship of the Mafia to the music business continues to be the subject of numerous books and innumerable conversations throughout the record industry. Both the popular music and crime culture of this country share their "other side of the tracks," "bad part of town," "outlaw" origins, and it is impossible to be in any part of the industry—distribution, talent management, retail, performance venues, etc.—without an instant awareness that there are "goodfellows" among the good guys and bad guys.

I was employed at Roulette Records when it was owned by Morris Levy, the single most notorious music executive ever identified with the underworld. Although we made many records which at any company would have been artistic, critical, and commercial successes, a few minutes in the office corridors or reception areas was all anyone needed to be aware that Roulette was a small subsidiary of a vast international mega-conglomerate that never filed with the Securities and Exchange Commission, and whose board of directors and shareholders met at clam bars in Brooklyn, Las Vegas, Naples, and Palermo.

Morris "Moishe" Levy was one of the nicest and most gentlemanly persons I ever worked with, despite his loud, gravelly voice and obvious, ever-present confederates. Inside the industry, Levy was widely known as the "official loan shark" and "Godfather" of the music business. He became incalculably wealthy by funding performers, music publishing, record manufacturing, and distribution companies, and others who experienced the cash-flow problems endemic to the industry. Instead of the traditionally usurious repayment installments such irregular loans would require, Levy demanded a controlling majority ownership in his "projects." He also got into a netherworld of life-perilous trouble when he refused to share increasingly larger portions of his ill-gotten wealth with *his* "bosses."

No matter how complex the details—which amazingly, he was always able to retain mentally—Morris was also known to scrupulously abide by any "handshake" commitment he made with his recording artists and executives such as myself. One of the most famous and acclaimed albums I ever produced—the Louis Armstrong/Duke Ellington meeting on Roulette Records—was very quickly organized because Joe Glazer, on behalf of Armstrong, and Duke, both of whom possessed the hardest noses in the music business, unhesitatingly accepted a Levy handshake instead of a contract to proceed with the recording.

In the late 1980s, and after many decades of relentless government attempts to jail him, Moishe Levy was finally convicted of extortion charges in a gangland kickback swindle that involved MCA Records "cut-outs" (unsold records excised from the current catalog). He died of liver cancer in May, 1990, at the age of 62 before he would serve his first day in prison. I will always remember him as one of the kindest, most warm-hearted, and classiest music men I had ever known.

The miasmal hoodlum atmosphere at Roulette Records was so heavily oppressive that it was often difficult for me to concentrate on the musical matters that were my direct and only responsibilities. In fairness, everyone was diligently circumspect about my "civilian" status and left me alone, even though everyday I felt I was improbably and inescapably trapped in a grade B gangster epic.

Among the clichéd central-casting characters who permanently populated the Roulette offices, I rather fondly recollect a domineering, silk-suited, pinky-ringed force of nature who was introduced to me as an executive with the Longshoremen's Union named "Dominick" (his last and/or real name was never, mercifully, revealed to me). Constantly engaged in hushed conversations with similarly attired associates in hallways and empty offices, his greatest pleasure seemed to be the preparation of pasta in the executive dining room kitchen at the label. Whenever I or another music executive did not have a luncheon appointment, Dominick would proudly and eagerly insist he cook a meal "better than that stuff out in the street." It was most obvious "The Dom" was a man whose generous offers could only be rejected at tremendous personal risk and regret by his invited recipients, and I was always delighted his invitations were those my appetite couldn't refuse.

Dominick was, however, no master chef by vocation as was repeatedly apparent on the most dramatically edgy days of all at the Roulette Records offices. On those frequent occasions, Morris Levy would surrender his imperially appointed suite on the executive floor, and Dominick would commandeer the premises to host intense discussions between imposingly menacing visitors and their ruffian retinues from the Bronx, Brooklyn, New Jersey, and boundaries beyond. I soon came to realize my interesting friend was the supreme court, judge, jury, and irrevocable arbiter who "mediated" all significant disputes involving those "gentlemen of respect" who were so visible in the corridors of Roulette Records the year I was employed there.

II

The unregulated riches that have always been abundant in show business made it elemental that the mob would be attracted to the industry. Performers of Italian descent were especially made to feel obligated to be "helped" by their brethren. Frank Sinatra, Dean Martin, Al Martino, Jimmy Roselli, Louis Prima, and Tony Bennett are a short part of a long list I had always heard of. I also remember stories about a respected talent agent named Lou Perry—who was Dean Martin's first manager—and the singer Vic

Damone (his parents, Mr. and Mrs. Farinola, named him Vito) being hung by their legs out of hotel penthouse windows when they did not "show respect" for demands to let "the boys" be involved.

As successful as Alan Dale (born Aldo Sigiamundi) was with me at Signature and then Coral Records, Dale insisted on a quiet life off-stage at home with his parents, and made no secret of his disdain for the frequent advances by the Mafia to manage him. One evening when he was headlining at the Latin Quarter in New York City, one of the top nightclubs in the country at the time, mafiosi hurled Alan down the entire length of the long staircase that was a landmark of the club. This highly visible incident nearly killed him, and with the mob now marking him as "trouble," his career was effectively killed as well.

Later, Johnny Desmond (real name: Giovanni Desimons), another successful Italian singer I produced at Decca/Coral—we had Top 10 hits together such as "The Yellow Rose of Texas," "Play the Hearts and Flowers (I Wanna Cry)," and even a cover version of "Sixteen Tons" which peaked on the charts at #17—recorded a song called "Women." Immediately after the session Desmond told me, "If we get this out right away, 'my people' can guarantee there will be an order of 50,000 records for jukeboxes."

In those days jukeboxes were a powerful and ubiquitous public medium for the promotion of popular music and a known mob-controlled industry. Before they were replaced years later by home and car radio receivers, with their easier, cheaper, and cost-effective access to even larger audiences, jukeboxes were essential to the success of any record.

Aware of Johnny's affiliations, and the impressive quantity of records involved, I rushed out "Women," which instantaneously became a jukebox smash and a Top 10 hit for all of us, staying on the pop charts for two months.

Soon after, a similar circumstance chillingly entailed the same maneuvering and initial sales figure. I was approached by the famed music industry (and Morris Levy's) attorney Halsey Cowan, who informed me that he just had lunch with Robert Costello, Frank's brother, who had confirmed a guaranteed order of 50,000 records from jukebox operators for any song by the girl vocalist, whose name I have long and gratefully forgotten, on a demonstration record that he requested I listen to. "Important people" were "interested" that Coral Records sign and record this singer. (I was certainly aware that, if I accepted this overture, it could open the door for the mob to dictate who and what a major label recorded.)

Luckily, the demo was so blatantly amateurish and the lady unmistakably untalented, I was able to say, "Halsey, she shouldn't sing; she should get married and have children," and survive a refusal to those "important people" without any of the calamitous repercussions I was becoming only too aware were possible.

14 Duke's Place

THROUGHOUT the years, I have been blessed with considerable luck, and the three best friends of my lifetime: my wife Teresa, my son Bobby Jr., and Duke Ellington.

Of all the jazz musicians I have known and worked with, it was Duke who made the deepest impression on me. When I was a kid and started to listen to music, I idolized him. I'd go out many nights alone, all by myself, to be around music, and somewhere along the line I was listening to Ellington night after night, and he spotted me, and we got talking. It was a natural thing, and it developed into a terrific friendship.

He was a worldly sophisticate of extraordinary refinement and intellectuality, an authentic musical genius, and unceasingly elegant in his wit, modesty, and audaciousness. (When, toward the end of his life, and to the outraged protests of the African-American community, he was again passed over for a richly deserved Pulitzer Prize, Ellington memorably replied, "Fate is being kind to me. Fate doesn't want me to be famous too young.")

And I was always flattered by him. When I was in the Coast Guard during the Second World War, Duke agreed to appear with his entire orchestra at the base as a favor to me. Actually, it was on that occasion an incident occurred which illustrated Duke's superb sense of delicious irony and complete lack of pretense, although he was at all times fully aware of who he was and what he singularly represented to people all over the world.

The band had gone ahead in the bus, and to my delight I was anointed as Mr. Ellington's personal chauffeur for the day. As the performance was, unusually for Duke in that sleep-a-day era, an afternoon event, he was predictably late getting started, and once he was regally ensconced in my auto, I decided to ignore all legal speed limits to get us to the base in time for the concert.

Inevitably, a motorcycle cop, seeing a car occupied by a large black man and driven by a white boy in a sailor suit streaking along the parkway at a speed many miles in excess of every conspicuously posted restriction, stopped us. "You don't understand, officer," I proudly announced, "I've got the great Duke Ellington in my car and we're late for his concert at Manhattan Beach!"

The cop was clearly unimpressed. "Listen, you squirt, I should arrest you for going that fast, and I don't give a fuck who you have in the car, here is your ticket"— a declaration immediately punctuated by several loud whoops and guffaws that erupted from my esteemed and absolutely delighted passenger.

Somehow we got there just as the band had started to play, and for the rest of the afternoon, as well as weeks after, every time he saw me Duke could not resist to mischievously admonish, "Pay the ticket, you still have the ticket, don't you? Why don't you pay the ticket?"

Also, whenever he saw me, alone in my sailor suit, at a club where he was appearing, he would ask his common-law wife Edie, or another attractive young lady, to sit with me, gleefully instructing them, "Make sure he doesn't fall off the stool."

Many times, after listening to his orchestra all night, I would be favored with the opportunity to drive him home, or to an occasional liaison. Duke would somehow squeeze his large bulk into the tiny Volkswagen I had then, and off we went. As regularly, when we arrived at his Edgecomb Avenue residence in the Sugar Hill section of Harlem, we would sit for hours in my car until dawn while Duke gossiped about the music business and discussed the growth of jazz and how important his musical heroes—Jelly Roll Morton, Willie "The Lion" Smith, James P. Johnson, and Fats Waller—were to the music and his own artistic development.

These euphoric journeys abruptly ended the night I drank
more than usual and, while driving Duke to his nocturnal assigna-
tion, hit a bus stop stanchion. Duke said, only half-kiddingly, "Hey,
man, don't ask me to drive with you anymore!"

Our fondness for each other survived that mishap, and, as the
years went by, I also got to know his family, and we all became
friends. Duke always seemed to trust me, and we did many things
together, socially and professionally. He and I became very close
and would do anything for each other.

Once, in half an hour, he wrote a song especially for me titled
"Blue Piano" that has since been recorded by many musicians. He
also allowed lyrics written by my friends and I to be added
to his timeless composition "C-Jam Blues," which then became
the popular song standard "Duke's Place." And whenever I wanted
to record him with anyone, whether it was with Louis Armstrong,
my wife Teresa Brewer, Coleman Hawkins, or John Coltrane,
I would just tell Ellington my idea, and he'd say, "Let's do it,
let's go!"

Soon after I first became friendly with Duke, when my friend
Dan Priest and I were publishing our neophytic *Jazz* magazine and
came out with an "Ellingtonia" issue, we got the bright idea to sell
copies in front of Carnegie Hall on a night Ellington was to perform
one of his concerts at this sacred musical shrine. An official came
out and told us, "You kids can't hawk magazines in front of Carnegie Hall." I went backstage and found Duke to explain my problem
to him, and he went to the management and said, "Hey, you let
these lads sell their magazine or we do not appear tonight."
Thanks to the Dukal amusement at our enterprise, our tyro tribune
was a huge success that particular evening.

An insouciantly charismatic, imperious, and accomplished
musician who was perhaps the first internationally visible African-
American celebrity to whom the term "Renaissance person" could
be applied, Ellington was a potent anomaly to the culture that
produced him as well as the current social and artistic milieu he
embraced and influenced. Duke was a black jazz musician who
inexorably demolished every stereotype that could attach to him,
and perhaps that is why, throughout his life, controversy followed
him like a parasitic relative.

One of the two times I was drawn into the contention that
persistently swirled around my friend involved our special issue of

Jazz called "Ellingtonia" that was entirely dedicated to his music. "Ellingtonia" included an article by John Hammond, "Is the Duke Deserting Jazz?," that criticized Ellington for composing large-scale works (instead of only the usual 32-bar songs) and increasingly appearing in major classical music concert venues such as Carnegie Hall (rather than nightclubs and dance halls — an irony, since Hammond's discovery Benny Goodman had opened up Carnegie Hall to jazz years earlier and, for that matter, had continued to commission modern compositions for performances with symphony orchestras. Perhaps John was being territorial—as common then as now in all arenas of aesthetic politics.)

Another essay in that same issue, "A Rebuttal of Hammond," was a vehement refutation of the Hammond piece written by Leonard Feather—by then a well-respected jazz critic as well as, in those days of exceedingly less rigorous journalistic standards, Duke Ellington's press agent. Also, Jake Trussell Jr., contributed a commentary titled "Ellington Hits the Top, and the Bottom" that agreed with John Hammond and traced Duke's "artistic deterioration" to both the various personnel changes in the Duke Ellington Orchestra—which always were, in fact, a frequent occurrence in the big band business, and his increased interest in concert settings. (In the article, Trussell castigated Carnegie Hall as "the sacrosanct, hypocritical hideout of everything and everybody that hates jazz music.")

As was typical of the explosively opinionated emotional climate that preoccupied the jazz world of the period, Jake Trussell Jr., next appeared in the pages of *Jazz* two months later with "In Defense of Hammond," in actuality a raging vilification of Leonard Feather, along with my exercise of an editor's prerogative, "The Case of Jazz Music," which attempted to mediate between the extreme views this squabble had generated. A few months after this exchange, the publication hosted a Jim Weaver fusillade on the subject titled "Jazz and Ellingtonia," which attacked *all* of the previous writers.

To further indicate the prevailing sentiments, naïveté, and confusion of that time and era, and my personal overview of the music and the Ellington dispute (as well as my sometimes awkward grammar), let me reproduce "The Case of Jazz Music" exactly as it first appeared in the July 1943 issue of *Jazz*:

I intended to title this article "The Case of Duke Ellington" and add
my bit to the current Ellington controversy being carried on in *Jazz*.
In this article, besides stating my viewpoints on the situation, I
intend also to add to some of the points in the articles written
by John Hammond, Leonard Feather and Jake Trussell in the last
issue. I will do all of this and more, because I feel that the Ellington discussion leads up to factors important to the good of Jazz
music. . . .

Critics have been preaching about Duke's music for years and
now that he has been recognized as a great musician, composer
and arranger he has come down from Jazz Heaven to walk with us
mortals. This is just so much "hokum."

As far as I am concerned Duke Ellington was and always will
be the most powerful force in jazz music. He is an ingenious arranger, a brilliant composer and a pianist of talent. He has proven
over and over again that his orchestra has never lost freedom and
spontaneity, the essence of jazz. He writes with a feeling for jazz
and his musicians interpret the music with the same feeling. Duke
also leaves plenty of room for improvisation. However, in the past
few years it has become quite evident that Duke is filling his ambition to work in more extended orchestral forms. Many of his arrangements are definitely influenced by modern composers.

In years past Duke's band has always strived to present individual and ensemble performances that were innovating and yet
played in a true jazz style. But lately, many of Duke's arrangements
present a love of exaggerated coloring, tending toward a sort of
varied, over-rich layer cake of ideas and tones. I am afraid I have
no sympathetic appreciation of this type arrangement, for it is in
direct opposition to the fundamentals of jazz. I am inclined to agree
with John Hammond, when he states that Duke is drifting further
and further away from dance music. After all, jazz is dance music.
Leonard Feather complains that Hammond doesn't know how to
dance and no one cares to dance in Carnegie Hall. This may be
true, but Leonard also stated in his article, "Duke's music has gone
a little beyond the stage where it has to tickle the toes of a mob of
jitterbugs." That is just the point. True jazz must contain that beat,
and once it reaches the concert form, it is no longer dance music.
Jazz must be free and exciting; spontaneous and spirited. As a
musician, Duke merits the warmest commendation for trying to
better himself in the field of music, but let's not say *Black, Brown*

and Beige is a thrill that cannot be compared with anything else in jazz.

The conclusion I have come to concerning the present Ellington situation is that, by becoming more and more involved with music by the modern composers, Duke Ellington is slowly losing contact with the basic fundamentals of hot jazz.

In answering the articles by John Hammond and Leonard Feather I make the following sincere suggestions. It might be wise for Leonard Feather to listen to more healthy bursts of New Orleans music and for John Hammond to go and listen to Duke under proper conditions. I recently heard Duke's band play two one-nighters and can honestly say it was the most exciting music I have heard in many years. Duke's band can definitely "swing" and I feel that if John should happen to hear it when it is "swinging" he might change a few of his opinions.

I realize that Leonard Feather doesn't like the idea of John Hammond suggesting what men to use to an orchestra leader, but I am about to do the same thing, for I feel that all jazz enthusiasts have a right to make suggestions.

I agree with Jake Trussell [in the May 1943 issue] because I feel that many of the men who are leaving Duke cannot be replaced and can rightly be called a part of "Ellingtonia." Jake's article was humorously exaggerated, but he is correct about the effect the absence of these musicians will have on Duke's music. Cootie Williams, Rex Stewart, Barney Bigard, Jimmy Blanton and Otto Hardwicke are no longer with the band. Recently I spoke with Duke for an hour or more and he told me that when Otto Hardwicke left the band it would be necessary to re-write the entire lead sheet [i.e., lead alto parts], because, as Duke explained, he wrote the sheet for no one but Otto Hardwicke. It is not beyond the realm of possibility that Duke may be able to build new soloists out of the group he has at the present time. Time will tell.

It wouldn't be necessary to wait if Duke had picked exceptional musicians to replace his great soloists of years gone by. It would be impossible to replace Cootie and Rex, but a clarinet player with a fluid, effortless style could have been found to replace Bigard. At the present time Duke is using Jimmy Hamilton. Jimmy told me two years ago, while he was playing at the Village Vanguard and studying at the Juilliard School, that he was striving for a semi-classical tone and a technique comparable to Benny Goodman. Two years later I find that he has almost achieved his goal and he should fit into typical Ellington mood-numbers very well. Junior Ragland [Raglin] is not up to the past standards of Ellington bassists. Duke has added two sax players that shouldn't be anywhere near the great Ellington band. However, in the singing of Betty Roche, Duke has an ideal replacement for Ivie Anderson. Betty has a very definite feeling for the blues and her voice is strong enough to take her out of the Harlem jump singer category.

All this leads up to an important question: What constitutes the music we are talking and writing about?

Jazz music springs from folk music and still contains many of its qualities. It is spontaneous, full of improvisation. It is music that springs from the *soul* of musicians. It represents America: Negro spirituals, marches, Tin Pan Alley. It is living American music. It is hard music, beat out for hard dancing. It is free music. It is comparatively new and different. It is rough and exciting.

I feel that jazz must always contain many, if not all, of these fundamentals or it is not real jazz. Unfortunately, present-day musicians are forgetting this, and so are many of the critics. They feel that when Duke Ellington plays a "different" chord it is truly great; *Black, Brown and Beige* is a thrill that cannot be compared with anything else in jazz. *Black, Brown and Beige* is not true jazz. Louis Armstrong improvising the blues is jazz. Billie Holiday's singing is jazz. Why is it hard for musicians and critics to grasp the ideas that constitute real jazz?

"Technique! The very word is like the shriek of outraged art." At times how true this statement rings in my ears. Most musicians who think they are playing wonderful music are merely stressing technique and what they consider to be good tone. It seems that the wilder, louder and the more notes he can inject into a chorus make him a great artist. These musicians are too weak to try to create something of beauty; they are content to forget about music that was played in the past. If some present-day musicians would take the time to listen to the mighty jazzmen, who hadn't much theory of technique, they might find themselves in a new world of music. However, many of the new crop of so-called jazz musicians are just innately dull.

I do not want my readers to feel I am intimating that such musicians as Art Tatum should be placed in the above category. Definitely not. But why must musicians rave about Tatum's technique, which *is* extraordinary, when he plays so much *music*? Why must young trumpet players base their styles after Roy Eldridge when Louis Armstrong is still playing? I wonder if some of these youngsters ever knew that Joe Smith had more finish and subtlety than Eldridge will ever have? Smith had a marvelous tone, round and full. Joe would stand up in the last row of the old Fletcher Henderson band and improvise two choruses and you could hear a pin drop. Not, "Take another, Joe."

The fact that a musician can send his fingers down the keyboard in a few seconds without missing a note doesn't mean he is a great jazz musician. Because a musician plays the same descending figures on every chorus, combined with a brass-pipe tone, does not make him a great jazz musician either. Remember, it is all very interesting music, but ask yourself if it's jazz. Let's not follow in the footsteps of Panassie by falling in love with the hybrid in jazz music.

It might be wise to adhere to a few simple words of wisdom uttered by [trombonist] Benny Morton.

"Jazz is Negro music. It has a tempo that's been handed down for a generation. It's easy and it rocks. There's no need to blow hard. Relax. Close your eyes and improvise melodies of beauty. Jack Teagarden still plays jazz."

In retrospect—and one wonderful advantage with a memoir is that you can say "in retrospect" about anything you previously wrote or published—I was unforgivably overboard in my narrow statements about these Duke Ellington compositions. Maturity and my extreme good fortune in continuing a close relationship with Duke made me aware of the true worth of *Black, Brown and Beige* and his other extended, "serious" works for me. These Ellington masterpieces are also now accepted as among the immortal and most significant exemplars of soulful, classic jazz.

Several decades later I would become even more directly embroiled in an Ellingtonal dispute that, for a brief while, seemed to engage many of our friends in the jazz world.

A hack songwriter and relentless music industry wannabe named Don George (who never seemed to have anything better to do than hang on and around Duke Ellington) wrote a blatantly self-serving, error-filled, and falsifying book in the early 1980s titled *Sweet Man: The Real Duke Ellington* which purported to be a candid narrative of his "relationship" with Duke. Don George had, in fact, insinuated his abysmally trite lyrics onto the Ellington "riff" standard "I'm Beginning To See the Light" and lucked out with the loony (and equally trite) pop hit tune, "Yellow Rose of Texas." (To be fair, Duke was always graciously tolerant of anyone's excesses: mine, his musicians', friends', and even those of the fawning members of his persistent entourage.) Everyone in the business who had worked with Duke or was privileged to be considered his friend was infuriated that this grasping, self-promoting sycophant would contrive such a literary fraud to publicize himself.

One chapter was about the making of Duke's 1973 album with my wife Teresa Brewer that I produced—an album we and Duke regarded as among the favorite recordings of our individual careers. Don George's depiction of it was so filled with inaccuracies, offensive innuendo, and disrespect for Duke (who had already died when the George book was published), I was forced to write a letter to *Jazz Times,* then and now a leading monthly publication, to correct the record.

Among the many falsities George alleged were statements of my "desperation" to have Teresa record with Duke Ellington; how I kept telephoning and begging Don George to help bring the album about (and if not for his exclusive exertions, the recording would not have materialized); that I was "cold, almost silent" during the recording except when I fired one of the "habitually drunk" musicians; and finally, that I cheated Duke out of any money he could have realized from the project.

Aside from my factual refutations of these bogus claims, my letter also described the longstanding working and personal relationship Duke Ellington and I enjoyed. My response immediately precipitated a considerable response in jazz periodicals from fans and respected industry professionals who were friends of both Ellington and myself. A few months after, Don George had been thoroughly discredited and his book consigned to the ignominious oblivion that was its merited destiny.

My letter, as printed in the March 1982 issue of *Jazz Times* (and after I briefly summarized my many years of friendship and collaboration with Duke), stated:

Now to get to the nitty-gritty of the present situation and as to why I am upset! Duke was such a dear friend and Don George, who

wrote this damn book, wrote two pages that describe something
that really never happened.

First, let me say that I never considered Don George one of the
great lyric writers of our time, even though he wrote a few things
with Ellington. The album by Teresa Brewer and Duke Ellington
came about in simple fashion. Teresa and I were at the Rainbow
Grill one night and Duke came over to our table and jokingly said
to Teresa, "I see you made an album with Count Basie. Are you
going to do one with my band?" I answered for her by saying, "Yes,
of course, let's do it." But knowing Duke, I knew I would have to
keep after him, that his remark didn't mean we would be going into
the studio the next day. So when I saw [Duke's son] Mercer I said,
"When you are talking to Duke, remind him that we want to do the
session with Teresa." Next I met Don George—he's always walking
up and down Seventh Avenue—and said, "Don, I know you've written a few things with Duke. When you talk to him, mention that
we're going to do this album." He loved the idea and said, "I assume
you'll be doing a lot of my songs." I replied, "It's only natural that
three or four of the songs will be ones you've written with him. The
rest will be older things, like *Mood Indigo* and *It Don't Mean a
Thing*." I think he was annoyed that the whole album didn't consist
of Ellington-George songs. There's one song in the album call *Poco
Mucho*, which is a trite, almost trashy little thing. Melodically it
grooves along pretty good, but even Duke referred to it as a "ditty."
Don had put the lyrics to it and pleaded with me to include it in the
album. I did so as a favor, with no strings attached. You try to be a
nice guy to a lot of people and . . . well, I think he was annoyed that
all the songs he wrote with Duke were not the basis for the LP.

Duke and I prepared the album. I suggested we get Ernie Wilkins to write some charts and Duke agreed. I am shocked when
Don George writes, "We never got a dollar for the album." He was
not a producer of the album nor an artist on it. He was not entitled
to any money except compositional royalties. He is a songwriter. I
think *he* would be in for a shock if he talked to Cress Courtney,
Duke's longtime friend and manager (and, a rarity in the business,
a very honest and dedicated guy). It was Cress who really swung
the album for me. When he and I discussed the advance for Duke,
Cress said, "I think he would do it for fifteen thousand dollars."
Well, I had originally allocated that amount in advances over and
above all recording costs for Teresa and Duke, so when Cress sug-

gested $15,000, Teresa said, "Pay Duke—I don't need a nickel." So
the only performer who did not get an advance on the record was
Teresa Brewer. Where royalties are concerned, the album never
sold that well and never passed its cost, which, off the top of my
head, was around thirty-five to forty thousand dollars. It has since
been re-issued by Columbia, so there will probably now be some
royalties for Teresa and Duke. But in no way was there ever any
understanding or agreement that Don George would share in advances or royalties.

So far as all the desperate phone calls I am described as making
to Duke and George are concerned, the truth is that I was running
away from the latter, who kept hounding me about what songs to
record. When he started to call Teresa at home, with suggestions
for songs, we even considered changing our number.

The book also describes me as being "cold, almost silent
throughout the session." (There were three sessions.) I don't run
around like a flaming idiot. I've worked with over three hundred
featured artists, have recorded at least a thousand different musicians, and have put out literally thousands of recordings. Maybe
Don George had never been to a properly run record session before, but everything was under control. I was responsible for every
take and all final decisions. When we did *Mood Indigo*, Teresa
thought there should be another take, but I assured her that the
record happened to be one of the best she ever made. Duke came
up and said, "Bob is right, Teresa. That's a great, great record." And
it was really at that point that Duke seemed amazed by her ability.
"You have a God-given talent," he said. "You swing and that's what
it's all about." So what was really a joyous and great occasion became a big drag in Don George's description, which fitted his own
warped idea of what happened.

The way he writes of (Duke's star sidemen) Ray Nance and
Paul Gonsalves is, to my mind, contemptible. I have always been
known by musicians as a friend and I had known these two for
many, many years, long before Don George probably ever heard of
them. Ray did get a bit out of hand on the first date, but Mercer
Ellington and I arranged his removal. And what if Paul Gonsalves
did sometimes drink too much? I doubt if Paul's wife enjoyed reading in this book that her husband was "drunk as usual." In fact, he
was not even present on the first two dates, because he was in

Providence for his mother's funeral. He came by on the third to see Duke rather than with any intention of playing.

What particularly bothers me is that a reputable publisher like Putnam would permit such a book to be written without checking the facts. Of course, I wonder about the validity of the entire book. I think Don George dreamed up half the stuff that's in it. It glorifies Don as a close friend and confidant of Ellington and tries hard to give the impression they were always together, but I *never* saw Don with Duke till we made that album. Then the pages about Duke's birthday party at the White House give the impression Don was there, but he was not on the invitation list and he was not among those present. I know, too, that both Duke's sister and son refused to cooperate in the writing of this book.

There was more, which referred to Ellington's personal life that, as I also indicated in my letter, was nobody else's business then or any time since.

Duke was one of the best friends anyone could have. He was there for me whenever he was needed, and I was honored to be counted as among the good friends he could always rely on.

Duke Ellington was also the greatest American jazz composer/arranger that ever lived, and even though he did not have as many popular hit songs as Cole Porter or Irving Berlin, Ellington left an immeasurably more valuable artistic legacy. Unfortunately, and like John Coltrane and too many others, he was deprived during his lifetime of the increasingly substantial respect and serious attention that exists for him today.

15

They Write the Songs

I

ALONG with the immortal jazz musicians I have been fortunate enough to hear and work with, the great American songwriters whose works contained more than five notes and three chords were always my idols. These days, I am pleased to see that the rock musicians have finally discovered the great songsmiths that preceded them, and have begun to appreciate the invaluably rich legacy represented by these creative giants of the pre-rock eras.

Eventually, I became a songwriter myself, and while I could never expect to be compared to my heroes, I am certainly sensitive (I hope!) to the agonies of song creation and how, as these next examples illustrate, this frequently painful and ceaselessly mysterious activity can affect one's personality and relationships.

Johnny Mercer, one of the premier lyricists of American popular song, and I became very close, and, like myself, he was an enthusiastically unashamed drinker for many years. One evening during this time, flushed with success and sloshed to excess as a result of celebrating Johnny's hit song "Something's Gotta Give" that I had produced for the McGuire Sisters, both of us decided to visit the Copacabana, the famed New York City nightspot where they were appearing.

"Something's Gotta Give" had originally been written for Fred Astaire to sing in the film *Daddy Long Legs*. It was already a classic song, and as it was one of the rare occasions he was the author of the music in addition to lyrics, Mercer was particularly proud of the result.

A few words about the McGuires. Fine, decorous ladies from a large and nurturing Midwest family, they were discovered and completely controlled by Arthur Godfrey, who promoted them into one of the most famous sister acts in the history of the entertainment business. In environments away from the public such as television and recording studios, they were charming and charismatic. However, performing for live audiences was so uncomfortable for these innately shy singers that they would be as wooden as a forest of statues, and every moment of their act had to be precisely scripted.

Upon arrival at the Copa, record producer in tow, John regally (if somewhat woozily) entered the nightclub as royalty bequeathing his imperial aura to awed subjects. The Sisters had already started their set, and, along with much reverential fawning by the management, we were given a specially held front table. (Johnny Mercer was, after all and in any condition, true show business royalty.) When the moment came for the McGuires to perform their current huge hit, they were predictably incapable of adding an ad-libbed recognition of the author's presence to their preordained patter, and simply introduced the song by its title.

Mercer, in front of the stage and packed house, flew into an abrupt rage. Noisily knocking over his chair as he stood up in the middle of the number, he loudly and slurredly proclaimed to me and the rest of the room, "They see me sitting here, and they didn't even mention my name, that I wrote this song. When you see them again, tell them to go fuck themselves!" And then, in all his wrathful majesty, Mercer thundered out of the club, leaving assorted displaced tables, seats, and customers, disconcerted waiters and maitre d's, an orchestra who had stopped playing, McGuires who had instantly stopped singing, and a thoroughly mortified record producer in his royal wake.

The words of Johnny Mercer had stopped many a show, but never this way.

II

Another personal icon, and national treasure, was Hoagy Carmichael, who, as our careers converged, also became a close friend. Although I would hear stories of his gargantuan anger, with me he was an unfailingly warm, openhearted "good old boy" from Bloomington, Indiana, who shared his unconditional love of classic jazz as few ever would.

Hoagy began his professional life as a pianist with such jazz legends as Bix Beiderbecke, Eddie Condon, and Frankie Teschemacher, and was one of the originators of the "Chicago Style," as jazz historians invariably refer to that indelible era. Arguably, Carmichael's most imperishable masterworks such as "Star Dust" can be analyzed as sublime improvisations that were written down and then lyricized.

I had a house in New Rochelle and invited Hoagy to spend the weekend. We were sitting in the kitchen having coffee on a sunny Sunday morning when my doorbell rang. It was Larry Spier, a music publisher/promoter. While industry people would utilize methods of varying creativity—legitimate and illicit—to bring their music to the attention of record producers such as myself, Larry's presentation was unique and popular. His mother had a pie company named Hortense Spier, and Larry would romance record executives with Sunday morning visits and gifts of delectable pastries—a most welcome and refreshing approach in a cynical and corrupt era of rampant payola dollars mated with drugs and debaucheries.

I was always glad to see Larry and be a recipient of his tasty crusade, and I invited him in for a cup of coffee. At the sight of Larry Spier, Hoagy hurriedly left the kitchen, angrily motioning me to follow.

Livid and wild-eyed, Hoagy left no room for discussion: "I hate that bastard. If he stays here a few minutes longer, I'm leaving. I'll walk down the street and just keep going."

It took a few very uncomfortable minutes to coax Larry to leave—without, of course, telling him of the crisis his presence had ignited—but too many for my now infuriated weekend guest, who commanded me to drive him to the railroad station for the next train back to Manhattan as promptly as possible.

Friends have ways to let you know that a past incident is too inflammable to ever talk about, and I could never find out what had caused this irreparable fission between "the pie man" and Hoagy Carmichael, one of the kindest, most effusively genial persons I had ever met.

III

The longest conversation I ever had with Irving Berlin was in a men's room.

Part of my responsibilities as the East Coast head of artists and repertoire for Decca/Coral Records was to visit the traditional New Haven previews of major musical comedies in the weeks prior to the Broadway opening night. I would then advise the company whether the show was a suitable event, commercially and artistically, for us to record as an original cast album. These recordings were a much more popular and cost effective genre than now, and all the record companies would fiercely compete for the album rights.

On this occasion, the show was Berlin's *Annie Get Your Gun,* starring the reigning diva of Broadway, Ethel Merman. About to become one of the supernally defining landmarks of American theater, the musical, produced by Rodgers and Hammerstein, directed by Josh Logan, and with a book by Dorothy and Herbert Fields, contained that luminous score with such (soon to be) enduring song classics as "Doin' What Comes Natur'lly," "The Girl That I Marry," "I Got the Sun in the Morning," "They Say It's Wonderful," "Anything You Can Do," and (next to Berlin's "White Christmas") the most popular popular song ever written, "There's No Business Like Show Business."

This *Annie* was among the most eagerly anticipated theatrical events of the decade, and that particular evening it seemed that the entire elite of the New York theater community had journeyed to New Haven to witness history being made.

As the overture began, professional that I was, I went to the theater lavatory sure no one would be there and I could return to my seat in time for the curtain to rise. There was one other individual nervously seated in the bathroom, as surprised as I was to see he was not alone, with the unmistakable horn-rimmed glasses, severe black hair, and bantam height known to countless music lovers everywhere.

Me (*shocked*): "Mr. Berlin, the show's starting."

He (*irritated*): "I know. I can't stand it. I never go near an opening, but I'm stuck here, and I'm going to sit in here until the show's over and then see what everyone thinks. Please leave me alone."

Decca recorded the original cast album—still a classic and permanent bestseller—and Irving Berlin spent one of the most momentous nights in the history of American musical theater in a men's room in New Haven.

IV

Great Songwriters of Our Time was an album series I devised and pushed through at Decca Records when I was still at the helm of their Coral label subsidiary. I had convinced my corporate overseers this would be as inexpensive as it was sure to have wide appeal. The idea was to record noted popular song composers singing their own personal favorite creations while accompanying themselves on piano.

The tunesmiths involved were J. Fred Coots (he wrote "You Go to My Head," "Santa Claus Is Coming to Town," "Love Letters in the Sand," "For All We Know," etc.), Arthur Schwartz ("Dancing in the Dark," "Alone Together," "You and the Night and the Music," "Something To Remember You By," "That's Entertainment," etc., etc., etc.), Harold Rome ("South America, Take It Away," the Broadway musicals *Wish You Were Here, Fanny*, and *I Can Get It for You Wholesale*—Barbra Streisand's first show and big break), and Bob Merrill (the shows *Take Me Along, Carnival* with the hit song "Love Makes the World Go Round," and the lyrics—Jule Styne wrote the music—to *Funny Girl*, which was Streisand's bigger break!). Actually, Bob Merrill couldn't play the piano so he accompanied himself on a toy xylophone.

Another honored composer who had also agreed to make one

of the albums was Victor Young, who unfortunately passed away before we started to record. Less regrettably, he was spared the indignity of an association with one of the more unsuccessful album series in music business history decades before the Great American Songbook found its substantial, loyal audience.

I particularly liked the Arthur Schwartz effort, which was an intimate and charming example of popular songwriting and performance at its most sublime, and, years later, I was pleasantly reminded of our unsung and unsold collaboration.

My wife, Teresa Brewer, and I were vacationing at a resort in the south of France, and one afternoon I saw Arthur at the poolside with his wife. I waved. He waved. And although I could tell he did not quite recognize me, I sent over drinks and waved again. Later that evening, when Teresa and I returned to our room after dinner, we were greeted with two brandies on our bedroom table and a note that said simply, "Thank you, Arthur Schwartz."

When we returned to the States, and now fondly re-energized to again work with Schwartz, I realized I had no way to communicate directly with him (I never had any contact with him after that day of poolside waves and nightside brandies). My only recourse was to send a letter to his son, Jonathan Schwartz, an inexorably unashamed self-promoter who, based on the esteem of his father, had established himself as a highly rated *American Songbook* authority and "personality" on New York City's classic pop music radio station, WNEW-AM—the renowned home of William B. Williams, Martin Block, "The Make Believe Ballroom," "The Milkman's Matinee," etc. Also, and because of his media status, Jonathan was now an aggressively frequent, anguishedly undistinguished cabaret performer.

My note described my past history with Arthur, along with the suggestion that Jonathan and I get together to co-produce for MCA the reissue of his father's *Great Songwriters of Our Time* album. Jonathan could also write the liner notes, and I was positive this classic recording would be a delightful and important musical event on compact disc for the now burgeoning audience who appreciated geniuses such as Porter, Gershwin, Carmichael, Berlin, and Arthur Schwartz.

Zip. Nothing. No reply. I was completely at a loss to fathom how a son could not respond to a letter about his father, and an obviously advantageous situation for everyone involved. This mystery only deepened some years later, during the only other time I was ever with Arthur Schwartz before he died.

It was during one of Jonathan Schwartz's semi-annual displays of arrogant narcissism at Michael's Pub, a nightclub/restaurant on the east side of Manhattan that was pleased to barter its live performance schedule for the ubiquity of free radio plugs a Schwartz appearance made inevitable. (Michael's Pub has since become internationally famous for the regular Monday night appearances of the Dixieland band led by comedian and film-maker cum-clarinetist Woody Allen.) This night, ostensibly billed as "A Tribute to Arthur Schwartz," had seemingly compelled the entire "middle-of-the-road" music community of New York to attend, I believe, even more out of obligation to the relentless broadcast "clout" of Jonathan Schwartz than to the considerable affection felt for his father.

Arthur remembered me this time and delightedly welcomed me to sit with him. For the rest of the evening, however, he seemed to become so embarrassed at the syrupy praise his son on stage directed at him, he could barely speak. And what could have been a happy reunion was a sad ending instead.

This, incidentally, was during the time I recall Jonathan was taken off the air at WNEW for his consistently negative on-air musical critiques of Frank Sinatra, the single most important artist programmed on the station, and best friend of William B. Williams, the single most important broadcaster at WNEW. Many weeks later, however, Jonathan got his airtime back after he publicly apologized to Sinatra, and privately entreated William B. to persuade "The Chairman of the Board"—Williams's famously endearing description of his friend Frank—to advise the station management that all was forgiven and Jonathan could be reinstated.

"Clout" in our industry is perpetually magical, mystical, and mythological. Once, soon after the Sinatra/Schwartz flap, when Teresa Brewer was headlining at a Waterloo Village concert in the New Jersey suburbs of the New York City metropolitan area, her record label bought a massive package of "30s and 60s" (thirty- and sixty-second commercials) on WNEW to support the appearance.

I was in my car with the radio on and heard Jonathan Schwartz, newly resurrected from Sinatracized oblivion to deific self-importance, give a most abusively indifferent delivery of the ad. Every listener must have known he hated to read it.

I pulled over to a roadside telephone booth and called Jim Lowe, who was the station's program director at the time. I begged him to have anyone else other than Jonathan do the commercial, and when I was told that would create a "political" problem, I asked the spot not be aired at all. The ad schedule had been paid for, but it would be less costly to eliminate the commercials than have Jonathan the Grate make the advent of a Teresa Brewer concert synonymous with a new inoperable cancer.

Jim Lowe then begged me not to pull the advertisements: "It gets miserable around here when Jonathan is upset, and you have to understand, he takes great pride in his hates. Obviously he hates Teresa and, if you must know, he is very proud of the fact he will never play a Steve Lawrence or Eydie Gormé record. He hates them also—even though our listeners love them. Since he came back on the station, Jonathan *only* likes Sinatra!"

16 Trane Tracks

I

MORE than a quarter-century since John Coltrane passed away, many people still tell me his music plays in their minds and memories every waking hour. And it was the same for me a long time after his death, and certainly during his last six years, when I was privileged to be Trane's producer at the Impulse! record label.

I was an executive at ABC Records when the Impulse! jazz imprint was formed in 1961 by Creed Taylor, a magnificent and subsequently influential jazz producer. The corporation had decided it should have a pop label, ABC Paramount, and a separate jazz component by reason of the massive popularity enjoyed by Miles Davis (with John Coltrane) and Dave Brubeck, among others, that had continued to widen from the prior decade.

Creed promptly made several fine albums with J.J. Johnson and Kai Winding, Ray Charles, and, perhaps most significantly, *Out of the Cool* by Gil Evans and his orchestra and the Oliver Nelson masterpiece *Blues and the Abstract Truth*. He also signed John Coltrane to the richest record contract (next to Miles Davis) a jazz musician ever agreed to, and produced Coltrane's first Impulse! accomplishment, *Africa/Brass*, which featured John, Oliver Nelson orchestrations, and a large brass ensemble.

In 1962, after nine albums, Creed left the label, and I, the resident "jazz freak" record executive at the corporation, was tapped to run Impulse! with John Coltrane as my principal production responsibility.

As a result of his recordings and appearances (starting in 1955) with first the seminal Miles Davis groups and then the Thelonious Monk Quartet, and his own pathfinding albums for Prestige and Atlantic Records, John Coltrane was among the most popular, important, and critically recognized jazz musicians in the world when I became his producer. He was an unfailingly courteous and truly gentle man who, despite his rumored and regrettably real drug abuse problems, seemed to have achieved an impervious inner serenity that further took on a profound spirituality when he discussed, thought about, or played music.

Also, and typical of most musicians who meet a record company executive for the first time, he warily eyed me like an off-the-rack empty suit to be most judiciously tried on before any possible purchase could be contemplated. There had previously been polite hellos and handshakes in the ABC offices and at trade magazine photo opportunity occasions, but Trane and I began to know each other at our first record sessions.

The startling *Live at the Village Vanguard* album was his second for Impulse! and also his first live recording as a leader. In what I learned would be his usual custom, he was completely prepared and focused musically to the point he chose not to discuss material with his musicians, to preserve the spontaneity, nor, aside from titles and personnel, with his producer.

Looking back, I remember being more than a little nervous about the decision to record live for three nights at the landmark jazz club (which had been, when I was a rosy-cheeked, adolescent jazz lover twenty years earlier, among my preferred hangouts). I was somewhat heartened that Trane brought a superb quartet consisting of pianist McCoy Tyner, Reggie Workman on bass, and his boundlessly innovative drummer Elvin Jones. (This group was the precursor of the summital John Coltrane Quartet with bassist Jimmy Garrison—who was also present at the Vanguard to play on some titles—in place of Workman.)

Other reassuring participants included my old friend from many record sessions and from John Coltrane's Prestige days, optometrist-cum-master-sound-engineer Rudy Van Gelder, and one of the all-time ultimate confidence builders, Max Gordon, the hippest and most nurturing pixie in jazz history and the Village Vanguard's legendary owner. Coltrane also informed me that his latest discovery would be on several tunes and then introduced me to Eric Dolphy, who I soon realized was among the most astonishingly revelatory saxophonists I would ever hear.

Coltrane and I were at the most initial stage of our relationship at the Village Vanguard sessions. I had previously heard some of his records and was aware he was turning jazz around. But I still didn't know him. There had been no reason to meet beforehand, since it had been predetermined by contract, management, and Trane's own quietly firm assertions that we would just record whatever he chose to play.

As I recall it, although Coltrane was at first extremely withdrawn and reticent to talk about anything, our friendship began during those nights we recorded. Sometimes these things happen. I've grown up with musicians. I deal with musicians, and I guess I relate to them. In this instance, as the music progressed and Trane became comfortable with the surroundings and results, his natural warmth and friendliness surfaced, and we hit it off.

The incident from those nights that I most fondly remember was a "name that tune" situation which involved an untitled blues Trane had conceived for the album. John instructed McCoy Tyner to "lay out," and without piano accompaniment, this twelve-bar pattern based on a Debussy motif suddenly conflagrated into a nearly sixteen-minute explosion of every improvisational idea, pattern, permutation, transformation, and abstraction three torrentially virtuosic musicians could create between, beside, and outside themselves.

As if possessed by the vociferant punctuations of Elvin Jones's drumming and, it seemed, all the raging, elemental forces of nature, Coltrane became a fevered dervish, a constant motion that the postage-stamp stage at the Vanguard could barely contain as he furiously hurled his fabled, searing "sheets of sound" from his saxophone at the apocalyptic counterpoint ignited by his accompanists.

Physicists have long debated about the existence of a "big

bang." Without any question, the jazz equivalent occurred during that seismic quarter-hour. Everyone in the audience was mesmerized: I was so intensely puffing the pipe I smoked in those days it nearly broke in my mouth, and, more incredibly, Max Gordon stopped counting his receipts to look up and listen!

It was Rudy Van Gelder, our indefatigable engineer, who immediately perceived that an invaluable document was about to be lost forever. Instantly he became a frenzied blur that mirrored Trane's unpredictable, compulsive choreography, resolutely keeping a hand-held recording microphone in front of the saxophone, continuously and courageously climbing over chairs, tables, waiters, and customers to accurately capture every precious sound.

And as suddenly as it began, this musical mega-nova ended. All of us associated with the recording were thoroughly drained, exalted, exhausted, and incapable to think of a title that would match the experience we had shared. It was Rudy Van Gelder who came up with the name that summarized both the extraordinary music and his own efforts: "Chasin' the Trane."

A sound technician of undeniable brilliance, Rudy Van Gelder was a glorious eccentric as well. Passionate about his ornithological interests, all-star recording sessions in his Hackensack, New Jersey, studio and residence were often interrupted so he could immediately run out—dragging with him as many puzzled and disbelieving musicians as he could entice—to witness the arrivals of various birds who, it seemed, always made the Van Gelder acreage a requisite stopover in their seasonal migrations. Furthermore, Rudy was more obsessed with hygiene and cleanliness than anyone I ever knew. The most famous jazz pianists in the world would be visibly timid to play his studio concert grand for fear of an unknowing despoilment of its pristine condition, and instant banishment. Rudy would also wear white gloves whenever he—and, needless to say, never anyone else—handled microphones, as he believed despicable human skin oils would destroy sound quality.

Even if you were Miles Davis or a record company president, smoking was not permitted anywhere on the property, even though several nicotine addicts came up with quite creative methods to circumvent this edict. When Van Gelder was once absent and I couldn't stand it, I sneaked a smoke, and perverse destiny caused an errant ash to mark his control room carpet. I desperately (and believed successfully) labored for twenty minutes to eradicate all evidence of my sin, but it was the first thing Rudy angrily noticed upon his return. Our friendship survived that calamity, and I was later invited to experience the ultimate Van Gelderesque privilege: to sit in his prized and seldom-driven "custom automobile of Italian manufacture" that was enshrined in his impossibly spotless garage. Naturally, he insisted I take my shoes off before I got inside!

II

After the Vanguard sessions, my relationship with Trane deepened. He brought me to what in that period was the "new jazz" or "new black jazz." The critics were giving it all sorts of names. (Another one at the time was "avant garde," and at Impulse! we encouraged the musical frontier concept with our own logo phrase "The New Wave in Jazz.")

In addition to Eric Dolphy, Coltrane introduced me to other brilliant musicians on the cutting edges of jazz such as Pharaoh Sanders, Archie Shepp, Albert Ayler, Gato Barbieri, Charlie Haden, and John's wife, Alice. They all recorded with him, and I was able to sign and produce them on Impulse!. Trane would always call me and tip me off as to who was good and where they were playing. Pretty soon—and John Coltrane was responsible for a great deal of it—Impulse! was known everywhere as the most important record company for innovative jazz on the current scene.

Coltrane also started explaining to me what he was trying to do musically. Previously, and being an old-time swinger, all I could tell was that it sounded as though he was literally leaving the chords. At first, when Trane was improvising, it just didn't hit me right. But I had a hunch that part of it was that he felt he could go further. He would tell me he always felt restricted playing within the chord, staying within the basic harmonies of, say, a Porter or Gershwin song standard. Trane would say, "Who says there has to be a restriction on what you play?" And the more I listened, the more it sounded natural to me.

Coltrane was the first player who really took vamps, improvising endlessly on three chords. But those vamps were just devastating. They were truly exciting. And not just to musicians—as proven by the huge record sales his versions of standards like "My Favorite Things" and "Inch Worm" invariably generated.

I think my contribution with Trane was to let him record whenever he wanted to—even when the corporate power structure was opposed to it. I knew I had to record him, and somehow I would fight the company. I believe his contract called for two albums a year to be recorded and released. Well, hell, we recorded *six* albums a year. And I was always brought on the carpet because they couldn't understand why I was spending the money to record Coltrane, since we couldn't possibly put out all the records we were making. If I was out during the day they'd ask, "Where's Bob Thiele?" and someone would answer, "Well, he's in New Jersey recording John Coltrane again." It reached a point where I would record late at night so at least we'd have peace then, and no one in the company would know where I was. (It was around this time that Elvin Jones started to refer to me as "the fifth member of the John Coltrane Quartet.") I'd come in the morning and say, "Oh, I recorded John Coltrane last night." Everybody would go crazy, but then it was too late. Besides, I figured if they were going to fire me, at least I would have made that last record.

Several of the records I made with Coltrane at Impulse! are still among my favorites. Both *Ballads* and even *John Coltrane with Johnny Hartman* came about because of the jazz critics, most of whom to this day I cannot understand. They seem to take petty delight in building up an artist and then attacking him as an "establishment" figure so they could continue to appear to be supporting the next "newest," "creative," "deserving," "now," and "relevant" musician. This meant that they did not ever have to address those more substantive considerations that determine the innovative and classic in every art form.

At the peak of Trane's popularity, most of the writers and various music magazines came up with a new category for his music and the musicians he inspired: "Anti-jazz." We decided to straighten these guys out once and for all by showing them that John was as great and complete a jazz artist as we already knew, and it was one of the few times he accepted a producer's concept.

I suggested that we do an album of popular songs, all ballads. We went in and did *Ballads,* a beautiful album that featured Coltrane with McCoy Tyner, Jimmy Garrison, and Elvin Jones interpreting such gorgeous song standards as "You Don't Know What Love Is," "Nancy (with the Laughing Face)," "What's New?," etc.

Both as a musical document and personal statement, John loved recording that album, and immediately became anxious to do a follow-up. This time I recommended we add a vocalist, and Trane, who always wanted to record with a singer, chose as his collaborator an old comrade who was experiencing some hard times, the veteran balladeer Johnny Hartman. Aside from the generous friendly gesture, Coltrane considered Hartman's rich baritone and musicianly phrasing of lyrics to be the closest approximation of his saxophone sound. And as Trane had begun to conceive the *Ballads* sequel as two singers—one of them himself on saxophone—with the Tyner/Garrison/Jones rhythm section, his longtime chum was a perfect partner.

All three of us selected the lush ballads for the album which included "You Are Too Beautiful," "Autumn Serenade," "They Say It's Wonderful," and "My One and Only Love." Almost the entire album was done in one take for each song, but I remember Johnny Hartman being so transfixed by Coltrane's elevatingly radiant solo during the recording of "My One and Only Love" that he completely forgot to come back in for his vocal close. A second take would be necessary, and, as tape-splicing was still a prehistoric craft, all of us in the sound booth were heartsick that a classic performance would be lost. For that next take, however, a buoyantly unworried John Coltrane then created the timeless solo that saxophonists will continue to imitate for as long as the sound of jazz is heard.

III

Ballads, the subsequently eponymous *John Coltrane with Johnny Hartman*, and the magnificent album recorded between those two pairing John and Duke Ellington caused the jazz critics to speedily revise their misguided bleatings. From this time forward, and as Coltrane expanded his visions to reach even greater artistic heights with such "unconventional" jazz masterworks as *A Love Supreme* and *Ascension*, his deserved significance as a musical and cultural force began to be permanently recognized.

I still believe among the greatest albums that ever happened was when I got John's agreement to record with Duke Ellington. I have never stopped hearing from everyone else how dissimilar they were, and yet, from the start, I thought what they shared in common was even more compelling.

Both Trane and Duke were accomplished, visionary musicians. Throughout their lifetimes, they both revered and celebrated the jazz traditions on which all of their music was based while they individually expanded and redefined that treasured heritage. Neither of them wrote "tunes" but invariably "works in progress" for the specific musicians they employed, and these compositions would continue to evolve as the associations deepened. This helps to explain why they became band leaders as soon as it was possible for each of them to do so. It was also the reason they each maintained a stable personnel for much longer than was usual in the jazz profession, and, irrespective of financial or personal considerations, absolutely insisted on keeping their musical organizations together—always on tour and recording.

Furthermore, there had not been two more enthusiastic, open-minded, generous, and nurturing talent scouts in the history of jazz, and both the Ellington Orchestras and Coltrane groups were invaluable in hiring, developing and graduating many of the most important jazz stars. And, too, John had started to become as dear a friend as Duke had been since I was a teenager.

With all the similarities Duke and Trane shared that made me anxious to record them together, there in fact was one dramatic difference in their approaches to music which evaporated once the *Duke Ellington and John Coltrane* session had started and not too soon for me.

As Coltrane's record producer, I had grown very concerned with his escalating self-critical obsession in the studio. I never could figure it out. Perhaps it was an awareness of his rapidly increasing fame, or, correspondingly, an extreme sense of responsibility, as countless musicians would tell Coltrane how important his music was to them. Trane had become impossible to be pleased whenever we recorded. He would ask for one take after another, with each subsequent take inevitably less exciting and genuine than the previous attempt. Then, as depressed as the rest of us, Trane would throw his hands up and say, "We'll try this again some other time." Probably the first or second take was a gem, but by the time there were twenty takes, a jewel had become paste.

We frequently lost extraordinary music. And the increasing costs of wasted studio time and salaries were threatening to become a serious problem that would add to the already large trouble I was in with my superiors by recording more Coltrane than we could possibly release.

I was particularly frustrated. I always believed one of the most essential functions of a producer is to keep the enthusiasm level high and let the musicians know when we've captured something classic. For this music, the spontaneity is so important that, the more you record something, the more it's liable to go downhill. You have to capture it when it happens—that's jazz—and Coltrane was now in the studio for hours on end, recording pieces over and over again.

Of course, Ellington knew, from decades of experience making classic records in the studio, if you get it, save it. His style was to capture the musical essence as quickly as possible on recordings. Don't destroy it by playing it over again if it's there in one take. Even if there are some flaws in a solo or whatever, if it's exciting, go with it. I rarely had hidden agendas with any of the artists I was privileged to record, however, but I hoped that recording with Duke might help allay John's increasingly detrimental self-doubts in a recording studio. If anyone could accomplish a cure, it would be the masterful Ellington, and I was completely confident we would also make a great album.

It worked out beautifully. From the start, he and Coltrane got along as two old cronies who had shared a lifetime of enthusiasms. Duke had big ears and knew all about Trane and appreciated him. And coming from a different era, Duke could make *anybody* feel comfortable, and make them think they were his dearest friend.

The first tune was Duke's immortal "In a Sentimental Mood," and after an exquisite first—and I was sure never-to-be-bettered—take, I looked out from the booth into the studio at John and Ellington, and only Duke was smiling. I knew just what was going to happen. Coltrane was going to say, "Let's do it again," and Duke was going to say, "That's beautiful."

I felt so good about the take that I ran over to Coltrane and said, "John, that was it," and Duke, as if on cue from a prepared script (he was unaware of Trane's super-critical compulsions, which I had never discussed with him) immediately added, "Bob, you're absolutely right. Why play it again? You can't duplicate that feeling. This is it. John, don't ever do it again here!"

So Coltrane said to use it as the master, and went on with both his and Duke's current rhythm sections alternating on each tune, and, in mostly first takes, to record quickly one of the all-time classic jazz albums. (Some time later, Johnny Hodges, Ellington's legendary saxophone star whose signature solo feature with Duke was "In a Sentimental Mood," told me he thought Coltrane gave the song "the most beautiful interpretation I've ever heard.") Also, and most reassuringly, after this session Trane became significantly less preoccupied with disproportionate quests for perfection.

IV

Another album Trane and I made together that I will always be fond of is his wondrous *A Love Supreme,* his "gift to God." As was customary, all John told me was that he was ready to record another album with his Quartet and that I would hear the music— which was very important to him personally—along with everyone else at the session. Inspired by Coltrane's religious convictions, *A Love Supreme* was an extended four-movement work that took up both sides of a long-play record and was extraordinary as both a musical and marketing event.

Its length was practically unprecedented in that era. Jazz musicians were ever mindful that jazz radio stations, which were essential to record sales, demanded each song be an absolute maximum of five to six minutes long; and made albums that usually contained ten selections. Now Coltrane makes an album with just four tracks, none less than ten minutes each (you can imagine how my bosses screamed about this!) that so moved and magnetized people that it began to be uninterruptedly played on all the jazz as well as classical stations. John's sales were around 25,000 to 50,000 over a year's time—tremendous for a jazz album—and *A Love Supreme* became his best seller, going into six figures and awarded by *Down Beat* magazine as Record of the Year in both its reader's and international critics polls. It also broke the radio "time" barrier, and made it possible for many more extended jazz composi-

tions to be written, recorded and heard, and achieve sales success
from then on.

Ascension, as with *A Love Supreme,* was all his, and another
work of transcendent brilliance that Coltrane revealed only at the
record date. This time it was a single uninterrupted composition
the entire length of an LP and, to my surprise, used many extra
musicians such as Archie Shepp, Pharaoh Sanders, and trumpet
star Freddie Hubbard—in addition to the John Coltrane Quartet.

Trane produced lead sheets for the players, and very quickly
and precisely told everyone what he wanted them to do and where
they were going, and that the music was just going to be continuous.

The complete composition was recorded on one day in two
takes. When I had everything transferred to cassette, I sent both
takes to Coltrane to listen to. I don't remember which take he
chose, but, by accident, I put out the other one. I then decided,
with Trane's approval, to play a little game on the jazz community.
After we went through the initial press run, I switched the masters
to the other take, the one John had originally picked, and inscribed
"Edition Two" on the inside of the runout circle on the lacquer. It
took several months before collectors and musicians started to
realize a different take was now released and say, "What's going
on?" So, really, there are two versions of *Ascension* out there.

V

I knew Coltrane during the intense period in the 1960s when the
long overdue civil rights insurgency had produced militant movements in all the arts. A black musician with massive popularity and
worldwide critical attention, John had become a profound cultural
icon for his people.

Of course, during that time, a lot of musicians were militant. I
was into it myself, and thought I was helping the cause by getting
Impulse! involved with some of those people. Charlie Haden, who
Coltrane introduced me to, and his Liberation Music Orchestra—a
militant musical organization—was definitely inspired by that period as were Charles Mingus and the Coltrane discoveries such as
Archie Shepp and Albert Ayler.

In those days it got pretty hot and heavy. A lot of discussions.
I used to have meetings with LeRoi Jones (Amiri Baraka), Stanley

Crouch—who also recorded for me as a poet and drummer—and Frank Kofsky, a white critic at *Billboard* magazine. Eventually they all became my friends, but for the literary fraternity during that period the music of Coltrane and others like Mingus spoke for the black militancy movement. Most of the musicians, including Trane, really weren't thinking the same way as their (militant) brothers. LeRoi Jones could feel the music was militant, but Trane never felt that it was but wouldn't go out of his way to say so.

John Coltrane never was one of those "haters," anti-establishment people such as LeRoi Jones or the Black Panthers, for example. Everyone really read more into his music than they should have. They assumed too much. Coltrane was playing *music* and that was it. I never heard him say two words about social or economic problems. He gained freedom from all of that through his music. John was really a very down to earth, relaxed, quiet man. As Trane explored musical possibilities, he was really finding peace.

Because, however, of the Impulse! records John Coltrane made, Stanley Crouch and I used to argue for hours and hours about the black movement and the music. I once asked him to write liner notes for a Duke Ellington album called *My People,* a theater piece Duke had put together which was basically a short history of blacks in the United States. I told Stanley, "Before you write this, Duke Ellington is not a militant. Take it from there." And the liner notes opened with: "Bob Thiele says that Duke Ellington is not a militant."

Stanley was being very subtle. He agreed with me when I pointed out something people should have realized a long time before then. When you think of his titles, Ellington was, in his own quiet, elegant way, trying to show the greatness of black culture in this country: "Black and Tan Fantasy," "Black, Brown and Beige," "Creole Love Call," etc. There was a record Duke made in 1929 called "Black Beauty" that I picked up when I was a teenager. I thought it was named after the horse because there was nothing going on then about various social problems. It was Crouch who said to me, "Hey, even back then he meant what he said: Black is beautiful." Duke, in his way, was the first person to say that. (As a matter of fact, "Black Beauty" was a musical depiction of the entertainer Florence Mills, among the first of many Ellington would compose that would later include portraits of Bert Williams, Bill Robinson, Ella Fitzgerald, and Billy Strayhorn.)

At this stage, I had actually joined in the "new jazz" movement, and it was invigorating. I probably became more excited and creative because of the musicians Coltrane introduced to me and those attracted to the Impulse! label. It really helped me continue my career. I just became more and more impassioned about making records—all kinds of records. Most of the producers who were involved with jazz just came up to the curtain. In my way of thinking, they didn't really progress with the music in the way John Coltrane had helped me to realize was both my responsibility and fulfillment.

I've known other record producers who, even though they were my contemporaries, always looked at me like I was a nutcase. They could never understand how I can listen to a John Coltrane and the "strange new music" he created or why people buy it. Yet many of these fellows made some of the finest recordings in the history of classic jazz.

And they all know Louis Armstrong's and Fats Waller's simple and profound jazz aphorism: If you have to explain it, then you don't know what it is. There really is no way to explain it. I believe that jazz is somehow inbred in the American people as they grow up. All popular music stems from jazz. Gershwin and composers from his generation wrote the music they did because they heard jazz musicians. They went to wherever there was a Harlem and heard jazz bands. Songwriters are still creating rhythmic popular songs. The song "Margie" swings if John Coltrane, Coleman Hawkins, Pharaoh Sanders, or Wynton Marsalis plays it. All of American popular music conclusively had jazz roots. What can you say? Either you feel it or you don't.

17   On Impulse!

I

AT Impulse!, I was really feeling it. Aside from John Coltrane, who was previously under contract, I actually went out and signed 99 percent of the musicians we recorded, and I know I was *living* in the studio. Eventually I made around two hundred albums. Many times I would be in the studio for 24 hours. We'd record Dizzy Gillespie, then Albert Ayler; I'd sleep on the couch, and then we'd do Freddie Hubbard or Sonny Rollins.

What I attempted at Impulse! was to record new young players such as Gabor Szabo and Gato Barbieri, and put them under contract. But then I also needed to build a successful jazz label, so I would record Dizzy, Rollins, and Count Basie on a one-shot basis, make the one album, and the Europeans and Japanese would be impressed that Impulse! was a substantial company.

We recorded every important jazz musician who was available—either for a multi-album deal or a one-off. Quincy Jones, Stanley Turrentine, Basie, Trane's pianist McCoy Tyner in various settings, Benny Carter, the herculean Art Blakey Jazz Messengers edition with Lee Morgan, Wayne Shorter, Curtis Fuller, and Bobby

Timmons all made records for the company. Sonny Stitt recorded with both Paul Gonsalves from the Ellington Orchestra and Art Blakey; Freddie Hubbard did a date with Sun Ra's tenor saxophonist John Gilmore; and my friend Duke Ellington made a superb album with Coleman Hawkins.

I was also especially proud of the many recordings I was able to do with Oliver Nelson, the genius composer/arranger and saxophonist. Those albums included *Soundpieces,* a sequel, *Further Blues and the Abstract Truth,* and *Alfie,* the best-seller that paired Sonny Rollins with Oliver's arrangements of music from the soundtrack.

Charles Mingus was another great musician who contributed to the distinctive Impulse! cachet. Notwithstanding *his* militant cachet, and his famous mercurial personality, Mingus was an all-right guy. (Once, however, when an ABC Records payment due him was late, I remember I walked into my office one morning to find a note with language and style of delivery that was vintage Mingus. It read, "Where the fuck is my money? MINGUS," and had been embedded in my desk chair with a knife.) In the studio and musically, there were never any problems. He always wanted to record, and was after me to schedule dozens of sessions. But I was budgeted and trying to make as many records as I could with everyone. We did make one of his classic albums together when he organized a large ensemble of jazz stars for his extended composition, *Black Saint and the Sinner Lady,* and he subdued his characteristically volcanic persona to make another of my favorites, *Mingus Plays Piano,* a solo keyboard recital of standard ballads.

And, of course, there were the still-famous Impulse! record cover spines.

I remember I was at a Newport Jazz Festival, and a guy walked up to me and asked, "Aren't you Bob Thiele?" I replied, "Yeah." He said, "I want to thank you for making those spines orange and black." (I thought he was going to say, "Gee, those recordings are great.")

The record cover spines were not just a fluke; some careful thought went into the idea. We felt that, no matter what the music or cover was, the spine for every record on the same label should always be uniform. Although this was an unprecedented concept, we did the spines for two reasons: the consumers would always be able to spot them in their collections, and, more important, the dealers would know where to look for the Impulse! releases when they had them filed on their shelves. It was all by design, and very effective. Ever since, my labels always uniformly display a color-scheme identification on the packaging. It's Marketing 101, and I'm always surprised so few other record companies do this.

In every way at Impulse!, I was at the jazz apex. Producing the music I loved most of all with the musicians I admired. The entire catalog was selling well, and Impulse! was finding an ever-larger audience for the music in both its mainstream and "new jazz" configurations.

II

Ever mindful, however, of the vicissitudes of the record business and lessons learned in prior corporate associations, I made sure I remained a presence on the ABC Records pop side. I produced hits for artists such as Della Reese, Frankie Laine, and much later, with "What a Wonderful World," Louis Armstrong. As added insurance, I persuaded the corporate powers to give me a blues label. We named it Bluesway, and I was able to sign and record such giants of the idiom as B.B. King, Muddy Waters, T-Bone Walker, Eddie "Cleanhead" Vinson, and Joe Turner, and actually start a national blues revival.

At ABC, my relationship with Frankie Laine was particularly pleasing. Since the mid-1940s with "That's My Desire," he had a seemingly endless parade of Top 10 hits such as "That Lucky Old Sun," "Mule Train," "The Cry of the Wild Goose," "Jezebel," "Jealousy (Jalousie)," and the unforgettable theme from the equally enduring film classic, *High Noon.*

I had always enjoyed and admired his work, and when one of the promotion men at the label, who had caught a recent Laine appearance at a Long Island concert, suggested I meet him, I was exceedingly interested. An industry colossus, Frankie Laine's most recent successful label relationship had soured four years previously when his great producer, Mitch Miller, left Columbia Records to lucratively transform the entire United States into his personal choral group.

Frankie had not hit the Top 40 charts in ten years, and had

failed to find a satisfying affiliation at various major labels. 15 minutes after I met this talented and gracious gentleman, Frankie Laine had a new deal, producer, and home at ABC Records.

At that time, the entire country was enthralled with 1920s nostalgia music, and singing along with most of it as a result of Mitch Miller's highly prospering efforts. "Winchester Cathedral" was also big then, and I suggested to Frankie our first record should similarly be "Every Street's a Boulevard in Old New York," which, incidentally, was written by Bob Hilliard, who had penned "Moonlight Gambler," one of the biggest Laine hits, years before.

Although he was amenable to my other suggestions for the B side selection, he asked that I consider a tune from the 1920s he just happened to have with him that he first heard in a Nevada bar and had been trying to record for years. It was titled "I'll Take Care of Your Cares," and I instantly loved it.

After we recorded it, there was no question "Cares" would be the A side. Almost immediately, the record became the first of Frankie Laine's nine hit singles with the company, and his return to the Top 40 pop elite, staying in the Hot 100 for over two months. I also learned that "I'll Take Care of Your Cares," a song that grew to be beloved as an enduring hymn to the open-hearted spirit of American generosity, was, in fact, the adopted anthem of all the Las Vegas hookers, and private recordings of it had been stocked in jukeboxes throughout that city for years.

Despite the "base audience" for "Cares," the song established a sentimental formula for us that included other successes as "Making Memories" and "You Wanted Someone To Play With," both of which also enjoyed multi-month stays on the charts. We were a superlative partnership for ABC Records, and even when other responsibilities prevented me from continuing as his producer, Frankie kept making money and hits for the label.

Then Larry Newton, the president of ABC Records and the nemesis of "What a Wonderful World," blew it.

As public tastes invariably changed, and Frankie Laine was re-established as a major recording star, Frankie correctly perceived that he should and could present a diversity of material to his new and expanded audience. Again, it was exactly the same as with "What a Wonderful World" and Louis Armstrong. Newton, who classified record stars and producers as niggling menials whose opinions and feelings were irrelevant, irrationally and staunchly believed that, if you were "lucky enough" to have a hit, it invariably had to be duplicated on all subsequent recordings.

He refused to allow Frankie any latitude in the material he would record, and disagreements quickly escalated into contentious warfare. Newton, at best a petty, boorish despot of venomous crudity—I've never understood how people like this become presidents of *anything!*—then began to treat one of his most popular artists as an inferior subordinate to be abusively insulted at every opportunity, and so a brilliant association was ruined.

In Laine's own autobiography, *That Lucky Old Son,* he echoed precisely what everyone who worked with Larry Newton felt: "I make it a practice never to speak ill of anybody, especially publicly, but I must confess that I've seldom disliked a fellow human being as much as I did Larry Newton, the president of ABC. He's one of the few persons I can honestly say that I met and detested immediately."

III

One of the more storied events in show business history, which has been increasingly mythologized as an unceasing turbulence of emotional trauma for everyone involved, was in fact among the easiest record dates I ever did.

When I recorded Judy Garland at the Palace Theater in 1961, I remember most of all that her dressing room was filled with more bottles of pills than I ever saw in any one place, including any pharmacy I have ever visited. She was also constantly drinking, but those vials and piles of pills!

Judy's then current husband, Sid Luft, had arranged with Larry Newton to record a live album of her excitedly anticipated series of performances at New York City's Palace Theater that I would produce for ABC Records. These since-fabled appearances were the most heralded of Garland's continuing and always exciting "comebacks" to date, and had rapidly exceeded the proportions of a millennial religious experience for her obsessively rabid, numerous fans in the New York area.

From the time I was assigned to document this prestigious occasion through the entire recording, editing, mastering, and graphics process, and working with the largest production budget a stringently conservative (read: *very* cheap) Larry Newton had ever given me during my entire ABC tenure, I never heard from or saw Luft, any of Judy Garland's other representatives, her children, ex-husbands, or a single senior ABC Records executive. No one was ever available to discuss repertoire, orchestrations, stage logistics, scheduling, etc., and as Garland was thoroughly disinclined to address these issues, I produced this immensely important album without any consultation, hindrance, or encouragement from anyone associated with the project. (Note to young record producers: DO NOT TRY THIS at home or anywhere else. This time I was extremely lucky the production turned out so well. Usually, and always preferably, the more collaboration with other involved professionals you can avail yourself of, the better the creative and commercial results.)

Through all of the engagement, Garland was so unflappably calm and relaxed (and seemed only vaguely aware of the recording), that I mistook what I eventually realized was her completely sedated condition for paralyzed fear. I kept reassuring Judy at every opportunity that the orchestra was great, these were the best audiences I had ever seen, we were getting consistently stupendous results on tape at every show, this was destined to be a classic album (I was certainly correct about that!), to absolutely no palpable reaction from her. Her only response to anything I said or did was the benediction of an amused half-smile and regal nod of allowance when a photographer entered her dressing room and I fell to my knees pleading for a picture of us both with me kneeling next to her.

Of course—and the album is glorious proof—even on automatic pilot Judy Garland was by herself more electric than every illumined marquee on Broadway. At every performance, both her uncontrollably ecstatic followers and I were mesmerized by the ineluctable magic of her stage persona and extraordinary musical talents.

In truth, the only moments of concern during the recording were provided by Reice Hamel, our sound engineer, who was as much a legend in his craft as Garland was in her sphere of accomplishment.

One of the venerated pioneers of live location remote (non-studio) recording, Reice was an intensely energetic, gnome-like man with a maniacally distracted genius temperament that demanded every mechanical object in his presence had to be promptly dismantled and reconstructed in an improved model.

It was always with anxious dread that I would visit his remote truck—actually a small van rebuilt so many times it had become a completely unclassifiable vehicle—to hear playbacks or check balances. (This was years before the ultra-sophisticated, technological luxury liner recording studios on wheels remote vehicles would become as an industry standard.) To walk into the Hamel "hearse" was to be immediately submerged ankle-deep in a jumbled jungle of wires seemingly unattached to anything, and innumerable abstruse electronic gadgets, doodads, whatnots, and whatsits. You never knew if you were about to be electrocuted, suffocated, or simply lost forever.

How Reice unfailingly achieved the exquisitely radiant sound he was so famous for amid the morass of mechanical rubble that seemed to be his requisite work environment was a continuing marvel. It was always too frightening for me to be with him when we were recording. I rarely saw any equipment I could recognize, and he perpetually seemed to be redesigning some sort of military device or space projectile. Consequently, I was continually running between the Palace Theater basement and the remote truck out on 47th Street to check sound balances, and what was, in all other respects, the most serene production experience I ever had was also the most exhausting!

Reice Hamel was undoubtedly a superb engineer, but I would leave his van every time once again overcome with incredulity and relief that both the irreplaceable master tapes and I had not disappeared in that four-wheeled Calcutta, never to be seen or heard again.

IV

With all the success at Impulse!, and frequently because of it, the skirmishing never stopped. As jazz record sales—and especially the "New Black Music"/"Avant Garde" Impulse! releases—visibly increased, the critics, ever fearful that any significant public popularity for the music was a terminal threat to their fragile legitimacy, were in a constant, screeching panic.

When I wasn't producing, signing, or sleeping, I was writing letters to magazines contradicting the braying denunciations of their pundits. It seemed the better our records sold, the more letters I had to write.

Then, too, and despite the prolific sales, the endless bureaucratic assaults and inexplicable inanities remained constant. I wanted to record Pharaoh Sanders after he appeared on John Coltrane's *Selflessness* album, and it took me two months for the approval because no one wanted to spend any money. The first record was made for scale—I think Pharaoh got a few hundred dollars for the album. And until the record came out, it was the same tired tune: "What kind of crap is this? This isn't going to sell; it doesn't mean anything; it's a lot of junk; you can't dance to it; you can't listen to it"; ad infinitum/nauseam.

But after the record was released, I remember the president of the company saying, "Hey, did we sign that Pharaoh Sanders?" I replied, "No, you didn't want to." "Oh," he said, "sign him up now; he's hot, let's get him." (I also recall that Pharaoh's second album on Impulse!, *Karma,* was at the top of the *Billboard* jazz charts for 12 weeks.)

When I look back at such instances, I am reminded that the corporate bosses at record companies where I was employed— such as Randy Wood, Larry Newton at ABC, and the executives at Decca, where I worked for so many years—could never be classified as music people. With their inveterate salesman mentality, these guys could sell shoes, shoelaces, or ice skates. In fact, these guys could sell *anything*! At no time did they really seem interested in the music, and they were never concerned with making any recordings that might have some social significance. I always got a kick out of Sidney Goldberg, the supra-efficient sales manager at Decca, for example, who consistently referred to Bing Crosby's timeless recording of "White Christmas" by the catalog number. He would never say, "How did we do on Bing's 'White Christmas'?" It was invariably, "How did we do on 24685?"

The gap between the chief executive types and music people like myself will probably always endure. Without question, we need each other's priorities, but, since the time I started in the record business as a teenager, I really have never made a record anticipating how it would sell. Fortunately or unfortunately, I think almost every record I've made was done because I personally liked the music. It's as though I was making records for my own collection.

V

As the 1960s were drawing to a close, I could sense my days at ABC were at an increasingly inevitable end. Larry Newton and I would never respect each other to have a productive, congenial working relationship, and, as rock was now intransigently dominant in the record industry with its performers who mainly wrote and independently produced their own songs, A & R people like myself had suddenly become obsolete.

The era of creative production staffs at major labels was over, and artists and repertoire had become an essentially administrative position. Since I could remain and push papers around a desk for the current crop of record producers or become an independent myself, the choice was easy.

I resigned, and with an appropriated name from ancestral mythology and a distribution arrangement with RCA Records, started my Flying Dutchman jazz label. Louis Armstrong and Johnny Hodges became Flying Dutchman artists, and both my friend Oliver Nelson and my discovery Gato Barbieri followed me from Impulse! to join my new enterprise where they made many fine records. (I first heard Gato at my former label when I recorded the Liberation Music Orchestra of Charlie Haden—the bassist whom John Coltrane had brought to my attention.)

I also found Lonnie Liston Smith, a former Rahsaan Roland Kirk, Art Blakey, and Miles Davis sideman who was a superbly trained musician equally involved with mysticism, transcendental meditation, and the newly apparent electronic keyboard applications. Our first album, *Further Expansions,* was one of the largest selling jazz recordings of the '70s and is still credited as among the most influential progenitors of the jazz-fusion movement.

Since the Bluesway label that I previously begun at ABC Records had brought the art form, both culturally and commercially, to new elevations of public awareness and acceptance, I was able to profitably continue the resurgence of the music at Flying Dutchman with a new subsidiary label I named Blues Time. We recorded such respected masters (and Bluesway alumni) as Eddie "Cleanhead" Vinson and Joe Turner, and re-energized the careers of other blues legends including Otis Spann and John Lee Hooker.

Blues Time was obviously doing well because after a few months I received a letter from ABC threatening a lawsuit based on the claim that the value of their property, Bluesway, was jeopardized because the name of my new label so closely resembled it. I didn't even have to telephone my attorney. I just wrote back a letter in which I said, "Although I am only a high school graduate, I do know that the word 'blues' is a generic term that cannot be copyrighted and you cannot sue me for using the word in either a song title or company name." I never heard from them again.

To continue making pop records, I also had a label named Amsterdam, on which I recorded my soon-to-be wife Teresa Brewer, while I began to produce jazz dates with her and many of the musical legends who were my friends. These sessions paired Teresa with such giants as Earl "Fatha" Hines, Benny Carter, and Count Basie and soon compelled an entire new worldwide constituency of fans and critics to become her loyal supporters.

Commercially, I was also able to indulge my now ripened social consciousness with various projects. The intense revolutionary ferment ignited by the civil rights movement of the previous decade had now been supplanted with the even more extensive public agitation incited by the national trauma of Vietnam and an unrepentant, imperiously antagonistic Nixon administration.

With jazz backup and Rosko, a popular New York radio presence who combined his street-smart urban sensibility with a mellifluous suavity that presaged, by at least a generation, the announcers of the now-in-vogue "Quiet Storm" format, we recorded the impassioned essays of (then) *New York Post* columnist Pete Hamill in a series of ground-breaking albums. Collector items ever since, among the more prominent titles were *Massacre at My Lai* and a commentary on prison conditions, *A Night at Santa Rita.*

I purchased master recordings by Angela Davis and H. Rap Brown to release on Flying Dutchman during this period, and my combative good friend Stanley Crouch enlivened the proceedings with his memorably titled poetry album, *Ain't No Ambulances for Niggers Tonight,* which did not immeasurably cut into the record sales of the *Mary Poppins* soundtrack released at around the same time.

I was so involved with countercultural awareness, I had the

ingenious idea to release a Spiro Agnew comedy album. As the Vice President was regularly assailing student demonstrators and the media on his speechmaking forays to events sponsored by administration sympathizers, I substituted laughter for applause on the live recordings, so every statement of solemn outrage uttered by the Veep became a punchline.

Enthralled and emboldened with my own rebellious accomplishment, I then sent a copy of the finished product, now titled *The Spiro T. Agnew Comedy Album,* to President Nixon with a note that brazenly stated: "You may or may not want to listen to this. I just want to be on record that we are not trying to overthrow the country, and hope you might like to hear this album." The LP, which had obviously been listened to after it was delivered to 1600 Pennsylvania Avenue, was promptly returned with an unsigned reply on White House stationary that read: "You are absolutely correct. We do not feel the President wants to listen to this record."

At Flying Dutchman, I discovered (or was discovered by) the militant black poet Gil Scott-Heron. A book of poetry he had written arrived at my office minutes before he somehow burst in to tell me that anyone who recorded Jack Kerouac and John Coltrane "couldn't be all that bad," and the perfect producer to document his reading of his work on record. It was all I needed to hear from an authentic spokesman of the militant community, and I instantly said, "That sounds great, great idea."

We finally figured out that to do the first poetry album—which famously turned out to be *The Revolution Will Not Be Televised* — we would use some rhythmic accompaniment. Then it sort of developed that Gil could sing and write songs, so we followed his recording debut with albums dedicated to Billie Holiday and John Coltrane that started a long string of successes for us.

The 1980s brought my next affiliation, CBS Records, and a new label which I named (thanks to that long-ago introductory conversation with George Segal) Doctor Jazz. We released studio productions with, among numerous others, Lonnie Liston Smith, Gato Barbieri, Teresa Brewer (now a respected presence on the international jazz scene), and the ageless Stephane Grappelli, and brought out the critically acclaimed All Star Road Band series of essential unissued Duke Ellington live performances.

The contemporaneous arrival of the compact disc, with its digitally improved sonics and longer playing times, attracted a

vastly increased audience of record consumers. Reissues soon became a significantly larger factor for the music industry then ever
before. All of a sudden, I was repackaging my classic Signature
sessions—which I now owned—on Doctor Jazz, while also producing reissues of my prior recordings for CBS, MCA, and RCA, which
still owned the original album masters. I even found time to produce a new Grammy award-winner on MCA, *Blues For Coltrane,*
with many of the star musicians who had played with him and
were previously signed to Impulse!—whose catalog was now the
property of the MCA corporate colossus.

Then, when the Sony Corporation bought CBS Records, a
former Flying Dutchman changed the name of his Doctor Jazz
label to Red Baron and continued to release more new dates,
reissues, and the music I have loved ever since I was a teenager.

Looking back as well as forward, I can say it was and is
unparalleled fun. And of all the record companies I ran or owned,
I suppose Impulse! was the best-named. It represented a unique
time when we were able to record a huge amount of music and
wide range of musicians than would ever again be possible. Impulse! most clearly exemplified the philosophy that had always
guided a once-pubescent, novice jazz record producer who would
be incredibly lucky to run that particular label—a label that could
never exist again—for nine of the most exhilarating years of my
life.

Partial List of Artists Produced/Recorded by Bob Thiele

John Abercrombie, Don Adams, Pepper Adams, Spiro T. Agnew, Airto, Manny Albam, Rashied Ali, Dayton Allen, Henry "Red" Allen, Peter Allen, Steve Allen, Laurindo Almeida, Don Ameche, Cat Anderson, Ray Anderson, Louis Armstrong, Svend Asmussen, Patti Austin, Albert Ayler,

Pearl Bailey, Gabe Baltazar, Amiri Baraka (LeRoi Jones), Gato Barbieri, George Barnes, Kenny Barron, Gary Bartz, Count Basie, Billy Bauer, Joe Beck, Louis Bellson, Sasha Berland, Gene Bertoncini, Denzil Best, Barney Bigard, Hal Blaine, Art Blakey, Terence Blanchard, Ray Bloch, Arthur Blythe, Phil Bodner, Pat Boone, Johnny Bothwell, Patti Bown, Bobby Bradford, Ruby Braff, Teresa Brewer, Lawrence Brown, Les Brown, Marion Brown, Ray Brown, Carol Bruce, Bobby Bryant, Ray Bryant, Hiram Bullock, John Bunch, Johnny Burnette, Ralph Burns, Kenny Burrell, Artie Butler, Frank Butler, Don Butterfield, Jaki Byard, Don Byas, Billy Byers,

Jackie Cain, Red Callander, Cab Calloway, Conte Candoli, Larry Carlton, Harry Carney, Benny Carter, Ron Carter, George Cates, Serge Chaloff, Carol Channing, Stanley Clarke, Rod Cless, Jimmy

Cleveland, Myron Cohen, Al Cohn, Henry Coker, Richie Cole, Bill
Coleman, Ornette Coleman, Buddy Collette, Al "Jazzbo" Collins,
Burt Collins, Alice Coltrane, John Coltrane, Ravi Coltrane, Eddie
Condon, Chris Connor, Jackie Cooper, J. Fred Coots, Chick Corea,
Don Cornell, Warren Covington, Bob Crewe, The Crickets, Bob
Crosby, Stanley Crouch, Ronnie Cuber, King Curtis, Andrew
Cyrille,

Alan Dale, Bill Dana, Eddie Daniels, Julian Dash, Angela Davis,
Eddie "Lockjaw" Davis, Richard Davis, Milton Delugg, Johnny Des-
mond, Paul Desmond, Billy De Wolfe, Vic Dickenson, Sasha Distel,
Eric Dixon, Eric Dolphy, Ray Drummond, Cornell Dupree, George
Duvivier,

Allen Eager, Harry "Sweets" Edison, Duke Ellington, Mercer Ellington, Don Elliot, Herb Ellis, Ziggy Elman, Skinnay Ennis, Booker
Ervin, Pee Wee Erwin, Bill Evans,

Jon Faddis, Victor Feldman, Cy Feuer, Jerry Fielding, Tommy
Flanagan, Rhonda Fleming, Bob Florence, Fontaine Sisters, Al Fos-
ter, Alex Foster, Frank Foster, Pete Fountain, Peter Frampton,
Alan Freed, Bud Freeman, Curtis Fuller,

Slim Gaillard, Judy Garland, Erroll Garner, Jimmy Garrison, Eric
Gayles, Terry Gibbs, Dizzy Gillespie, John Gilmore, Hermione Gin-
gold, Tyree Glenn, Arthur Godfrey, Paul Gonsalves, Benny Good-
man, Eydie Gormé, Danny Gottlieb, Stephane Grappelli, Buddy
Greco, Freddie Green, Urbie Green, Al Grey, Don Grolnick, Joe
Guercio,

Bobby Hackett, Buddy Hackett, Charlie Haden & Liberation Music
Orchestra, Bob Haggart, Connie Haines, Chico Hamilton, Jeff Ham-
ilton, Jimmy Hamilton, Lionel Hampton, Roland Hanna, Roy Har-
grove, Tom Harrell, Pat Harrington, Jr., Benny Harris, Bill Harris,
Johnny Hartman, Coleman Hawkins, Peter Lind Hayes, Roy
Haynes, Mary Healy, Jimmy Heath, Percy Heath, Neal Hefti,
Woody Herman Orchestra, Eddie Heywood, John Hicks, Earl "Fa-
tha" Hines, Milt Hinton, Art Hodes, Chas Hodges, Johnny Hodges,
Buddy Holly, Richard Groove Holmes, John Lee Hooker, Freddie
Hubbard, Marjorie Hyams, Dick Hyman,

Chubby Jackson, Milt Jackson, Bud Jacobson, Ahmad Jamal, Harry James, Gus Johnson, J.J. Johnson, James P. Johnson, Osie Johnson, Plas Johnson, Pete Jolly, Elvin Jones, Hank Jones, Jo Jones, LeRoi Jones (Amiri Baraka), Quincy Jones, Thad Jones, Victor Jory, Dick Jurgens,

Steve Kahn, Max Kaminsky, Dick Katz, Connie Kay, Carol Kaye, Danny Kaye, Roger Kellaway, Joe Kennedy, Jr., Jack Kerouac, B.B. King, Rahsaan Roland Kirk, Rod Kless, Don Knotts, James Komack, Roy Kral, Gene Krupa, Steve Kuhn, Billy Kyle,

Frankie Laine, Ellis Larkins, Louise Lasser, Yusef Lateef, Arnie Lawrence, Eddie Lawrence, Steve Lawrence, Hubert Laws, Yank Lawson, Albert Lee, Lennon Sisters, Jay Leonhart, Rod Levitt, John Lewis, Mel Lewis, Victor Lewis, Liberace, Kirk Lightsey, Abby Lincoln, Melba Liston, Charles Lloyd, Johnny Long, Harry Lookofsky, Dorothy Loudon, Mundell Lowe, Lorna Luft,

Teo Macero, Henry Mancini, Johnny Mandel, Shelly Manne, Mickey Mantle, Steve Marcus, Charlie Mariano, Joe Marsala, Marty Marsala, Wynton Marsalis, Phil Mattson & The P.M. Singers, Cecil McBee, Christian McBride, Gary McFarland, Lou McGarity, The McGuire Sisters, Hal McKusick, Barbara McNair, Jimmy McPartland, Marian McPartland, Sheila McRae, Jayne Meadows, Mike Melvoin, Bob Merrill, Mills Brothers, Johnny Mince, Charles Mingus, Liza Minnelli, Bob Mintzer, Modern Jazz Quartet, Miff Mole, Toots Mondello, Carlos Montoya, Lee Morgan, Buddy Morrow, Benny Morton, Mtume, David Murray,

Ray Nance, Marty Napoleon, Lewis Nash, Oliver Nelson, Joe Newman, Lennie Niehaus, Harry Nilsson, Louis Nye,

Anita O'Day, Sy Oliver, Glenn Osser, Horace Ott, Frank Owens, Jimmy Owens,

Gene Page, Hot Lips Page, Earl Palmer, Jackie Paris, Sonny Payne, Nicholas Payton, Dave Peacock, Curtis Pegler, Bill Perkins, Oscar Pettiford, Flip Phillips, Harvey Phillips, Bucky Pizzarelli, Alice Playton, Jack Pleis, Al Porcino, Tom Poston, Seldon Powell, Bernard "Pretty" Purdie,

Boyd Raeburn, Don Redman, Joshua Redman, Della Reese, Debbie Reynolds, Jerome Richardson, Boomie Richman, Cyril Ritchard, Max Roach, Howard Roberts, Red Rodney, Sonny Rollins, Harold Rome, Rosko, Charlie Rouse, Ernie Royal, Roswell Rudd, Jane Russell, Pee Wee Russell, Ali Ryerson, Art Ryerson,

Sabicas, Edgar Sampson, Pharaoh Sanders, Arthur Schwartz, Bobby Scott, Shirley Scott, Tom Scott, Tony Scott, Gil Scott-Heron, Don Sebesky, George Segal, Bud Shank, Sonny Sharrock, Dick Shawn, Archie Shepp, Wayne Shorter, Valerie Simpson, Zoot Sims, Derek Smith, Donald Smith, George Harmonica Smith, Lonnie Liston Smith, Marvin "Smitty" Smith, Lew Soloff, Otis Spann, James Spaulding, Jess Stacy, Marvin Stamm, Kay Starr, Bill Stegmeyer, Maeretha Stewart, Slam Stewart, Sonny Stitt, Carl B. Stokes, Billy Strayhorn, Ralph Sutton, Gabor Szabo,

Horace Tapscott, Grady Tate, Martin Taylor, Clark Terry, Bob Thiele, Jr., Toots Thielemans, Leon Thomas, Kay Thompson, Steve Thornton, Bobby Timmons, Dimitri Tiomkin, Brian Torff, Mel Tormé, Dave Tough, Margaret Truman, Joe Turner, Stanley Turrentine, McCoy Tyner,

Dave Valentin, Joe Venuti, Eddie "Cleanhead" Vinson,

Mal Waldron, T-Bone Walker, Clara Ward, Kenny Washington, Muddy Waters, Julius Watkins, Bill Watrous, Francis Wayne, Ben Webster, Bobby Lyle Weiss, George David Weiss, Lawrence Welk, Dickie Wells, Frank Wess, George Wettling, Paul Whiteman, Bob Wilbur, Joe Wilder, Ernie Wilkins, Billy Williams, Cootie Williams, Pat Williams, Richard Williams, Jackie Wilson, Teddy Wilson, Britt Woodman, Phil Woods, Sam Woodyard, Reggie Workman, World's Greatest Jazz Band,

Lester Young, Snooky Young, Trummy Young, Henny Youngman.

Bob Thiele's
Ten Favorite
Self-Produced Albums

1. LOUIS ARMSTRONG & DUKE ELLINGTON
   (Roulette Records—presently distributed on Capitol/EMI)

2. JOHN COLTRANE: *A Love Supreme*
   (Impulse!—presently distributed on Impulse!/GRP)

3. COLEMAN HAWKINS/LESTER YOUNG: *Classic Tenors*
   (Signature Records—presently distributed by Sony Music)

4. DUKE ELLINGTON AND JOHN COLTRANE
   (Impulse!—presently distributed on Impulse!/GRP)

5. JOHN COLTRANE WITH JOHNNY HARTMAN
   (Impulse!—presently distributed on Impulse!/GRP)

6. DUKE ELLINGTON & TERESA BREWER: *It Don't Mean a Thing If It Ain't Got That Swing*
   (Columbia Records)

7. DAVID MURRAY: *Saxmen*
   (Red Baron Records)

8. LOUIS ARMSTRONG: *What a Wonderful World*
   (MCA Records)

9. JACKIE WILSON
   (Brunswick Records—presently distributed by MCA Records)

10. THE BOB THIELE COLLECTIVE: *Lion-Hearted*
    (Red Baron Records)

Bob Thiele's Ten All-Time Personal Favorite Albums

1. LOUIS ARMSTRONG: *Volume II—The Hot Fives and Hot Sevens*
   (Columbia Records)

2. JIMMIE LUNCEFORD: *Stomp It Off*
   (Decca Records—presently distributed by GRP/MCA)

3. DUKE ELLINGTON: *1940–1941*
   (BMG/RCA Records)

4. DUKE ELLINGTON: *Braggin' in Brass*
   (Epic/Portrait Records)

5. ELLA FITZGERALD: *Songbooks*
   (Verve/Polygram Records)

6. BILLIE HOLIDAY: *The Original Recordings*
   (Columbia Records)

7. BENNY CARTER: *Further Definitions*
   (Impulse!—presently distributed on Impulse!/GRP)

8. YANK LAWSON: *That's a Plenty*
   (Signature Records—presently distributed by Sony Music)

9. DUKE ELLINGTON AND COLEMAN HAWKINS
   (Impulse!—presently distributed on Impulse!/GRP)

10. THE BEATLES
    (1962–70 on Capitol Records)

Bob Thiele's Ten All-Time Favorite Single Jazz Selections

(There are twenty because I have so many favorites!)

1. LOUIS ARMSTRONG: *Potato Head Blues*
   (Columbia Records)

2. JELLY ROLL MORTON: *Sidewalk Blues*
   (BMG/RCA Records)

3. LOUIS ARMSTRONG: *What a Wonderful World*
   (MCA Records)

4. DUKE ELLINGTON: *Cotton Tail*
   (BMG/RCA Records)

5. BENNY GOODMAN: *Ridin' High*
   (Columbia Records)

6. BUNNY BERIGAN: *I Can't Get Started*
   (BMG/RCA Records)

7. COLEMAN HAWKINS: *Body and Soul*
   (BMG/RCA Records)

8. COLEMAN HAWKINS: *The Man I Love*
   (Signature Records—presently distributed by Sony Music)

9. BENNY CARTER: *More Than You Know*
   (Bluebird Records—presently distributed by BMG/RCA Records)

10. COUNT BASIE (with LESTER YOUNG): *Lester Leaps In*
    (Epic Records)

11. BILLIE HOLIDAY: *Strange Fruit*
    (Verve/Polygram Records)

12. DUKE ELLINGTON (with ELLA FITZGERALD): *Cotton Tail*
    (Verve/Polygram Records)

13. CHARLIE PARKER & DIZZY GILLESPIE: *Groovin' High*
    (Savoy Records—presently distributed by Denon Records)

14. DUKE ELLINGTON: *Diminuendo and Crescendo in Blue* (Newport Album)
    (Columbia Records)

15. WOODY HERMAN: *Apple Honey*
    (Columbia Records)

16. MILES DAVIS: *'Round About Midnight*
    (Columbia Records)

17. BESSIE SMITH: *Gimme a Pigfoot (and a Bottle of Beer)*
    (Columbia Records)

18. JOHN COLTRANE & JOHNNY HARTMAN: *My One and Only Love*
    (Impulse!—presently distributed on Impulse!/GRP)

19. CHARLES MINGUS: *Goodbye, Pork Pie Hat*
    (Atlantic Records)

20. DUKE ELLINGTON & TERESA BREWER: *Mood Indigo*
    (Columbia Records)

Acknowledgments

Some authors have the good fortune to be blessed with remarkable collaborators as their books evolve from initial, hesitant jottings to final publication.

First among many, Sheldon Meyer of Oxford University Press shared his fanciful vision of a Bob Thiele memoir with two incredulous music executives who had never thought to write anything more serious than an occasional record album liner note. Through his unfailingly gracious professionalism, encouragement, and patient counsel, Sheldon has been an indispensable guide on an adventure that began as a whimsical exercise and ended with these printed pages.

Two consummate seekers of the best words, Jeff Levenson and Stuart Troup, continue to expand my appreciation of the difference (with apologies to Gore Vidal) between processing and writing, and, from the beginning, Bunny Spitz helped to preserve those virtues of focus and clarity exemplified by my mentors.

Special thanks must also go to Steve Allen, Murray Deutch, and Frank Military, whose generosity of time and recollection assisted these efforts immeasurably, and to Deborah Siegel for her invaluable intelligence, instincts, and enthusiasm.

Finally, a few more words about Bob Thiele, the extraordinary subject of this volume. He has my deepest gratitude for the loyal friendship and support that made possible the opportunity to be associated with *What a Wonderful World* as well as many of the rewarding years I have enjoyed in a music industry enriched by his talents and indelible contributions.

BG

Sources

Dannen, Fredric: *Hit Men;* Vintage Press/1991.

Fox, Ted: *In the Groove: The Stories Behind the Great Recordings*; St. Martin's Press/1986.

George, Don: *Sweet Man: The Real Duke Ellington*; G.P. Putnam's Sons/ 1981.

Hasse, John Edward: *Beyond Category: The Life and Genius of Duke Ellington*; Simon & Schuster/1993.

Knoedelseder, William: *Stiffed: A True Story of MCA, the Music Business, and the Mafia;* HarperCollins/1993.

Laine, Frankie & Joseph F. Laredo: *That Lucky Old Son: The Autobiography of Frankie Laine*; Pathfinder Publishing of California/1993.

Milkowski, Bill: "Irons in the Fire" (article); *Pulse!*; June, 1990.

Rozzi, James: "Bob Thiele Once Again Flying Dutch" (article); *Coda*; September/October, 1992.

Ruhlmann, William: "Bob Thiele Produced Them All" (article); *Goldmine*; December, 1992.

Shipman, David: *Judy Garland: The Secret Life of an American Legend*; Hyperion/1992.

Thiele, Bob: "My Sweet Friend" (article); *Jazz Times*—March, 1982.

Thomas, J.C.: *Chasin' the Trane*; Doubleday & Company, Inc./1975.

Tosches, Nick: "Mafia a Go-Go: The Unwritten History of Rock 'n' Roll" (article); *Los Angeles Times*—May 25, 1993.

Tucker, Mark (editor): *The Duke Ellington Reader*; Oxford University Press/1993.
Whitburn, Joel: *Joel Whitburn's Top Pop Singles 1955–1990;* Record Research Inc./1991.

Index